LIVESTOCK SCIENCE

Edited by **Selim Sekkin**

Livestock Science

http://dx.doi.org/10.5772/62514
Edited by Selim Sekkin

Contributors

Nagamani Balagurusamy, Athumani Msalale Lupindu, Bruno Do Amaral Crispim, Joyce Oliveira, Alexandre Banari, Alexeia Barufatti Grisolia, Juliana Dos Santos Fernandes, Stephane Ingrand, Majid Hussain, Emine Baydan, Sukru Kirkan, Alfredo J. Escribano

CBS, Edition **2018**

Published by InTech

Janeza Trdine 9, 51000 Rijeka, Croatia

Publishing Process Manager Iva Simčić

Technical Editor SPi Global

Cover Designer InTech Design Team

Additional hard copies can be obtained from orders@intechopen.com

Livestock Science, Edited by Selim Sekkin
p. cm.
Print ISBN 978-953-51-2863-2
Online ISBN 978-953-51-2864-9

Contents

Preface VII

Chapter 1 **Securing Sustainable Livestock Production Systems in an Uncertain Economic Climate: Nurturing Flexibility and Resilience 1**
Stéphane Ingrand, Laura Astigarraga, Eduardo Chia, Xavier Coquil, Christophe David and Jean-Louis Fiorelli

Chapter 2 **Do Organic Livestock Farms Differ from Low-Input Conventional Ones? Insights Based on Beef Cattle in Southern Europe 19**
Alfredo J. Escribano, Paula Gaspar, Francisco J. Mesías and Miguel Escribano

Chapter 3 **Naturalized Breeds in Brazil: Reports on the Origin and Genetic Diversity of the Pantaneiro Sheep 37**
Bruno do Amaral Crispim, Alexandre Campos Banari, Joyce Azambuja de Oliveira, Juliana dos Santos Fernandes and Alexeia Barufatti Grisolia

Chapter 4 **Livestock Methane Emission: Microbial Ecology and Mitigation Strategies 51**
Marleny Garcia Lozano, Peña Garcia Yadira, Karen Abigail Avendaño Arellano, López Ortiz Carlos E. and Nagamani Balagurusamy

Chapter 5 **Public Health Aspect of Manure Management in Urban and Peri-Urban Livestock Farming in Developing Countries 73**
Athumani Msalale Lupindu

Chapter 6 **Lactation Responses toward Milk Indigenous Enzymes 91**
Aneela Hameed, Majid Hussain and Saeed Akhtar

Chapter 7 **Bacterial Tick-Borne Diseases of Livestock Animals 109**
Şükrü Kirkan, Göksel Erbaş and Uğur Parin

Chapter 8 **Contaminants in Animal Products 129**
Emine Baydan, Murat Kanbur, Emre Arslanbaş, Farah Gönül Aydın, Semra Gürbüz and Muhammed Yasin Tekeli

Preface

Livestock production has played a key role in the development of human civilization. There has been a great increase in the consumption of animal products. Development of animal-rearing techniques has steadily increased the productivity. The intention of agricultural activities planned by governments, food industry, producers, and consumers is to provide safety and adequate food by respecting the environment. Although many questions are being raised about the sustainability of the world's food-animal agricultural resources. We begin to understand the detrimental effects conventional agriculture can have on local and global environments, with more studies focusing on alternative agriculture studies that compare alternative practices to conventional ones.

It is a pleasure to introduce on behalf of all authors the *Livestock Science* book. This book does not pretend to be fully comprehensive, but we believe it does provide topics which are of central importance. We hope this book will stimulate discussion about sustainable livestock production that meets the long- and short-term goals of human food production. This book presents some in-depth reviews of selected topics in livestock science written by experts in their respective areas. This book is divided into eight chapters, consisting of topics in food-animal production systems, management of several animal products, health-threaten example by ticks in animals, and contaminants that may be found in animal foods.

We expect that the *Livestock Science* book will be of interest to a wide readership. We hope that a wide variety of scientists, researchers, and students may benefit from this book. We also recommend it to the general reader, who will find much of interest in these chapters.

Selim Sekkin, DVM, PhD
Associate Professor
Adnan Menderes University
Faculty of Veterinary Medicine
Pharmacology and Toxicology Department
Aydın, Turkey

Chapter 1

Securing Sustainable Livestock Production Systems in an Uncertain Economic Climate: Nurturing Flexibility and Resilience

Stéphane Ingrand, Laura Astigarraga, Eduardo Chia,
Xavier Coquil, Christophe David and
Jean-Louis Fiorelli

Additional information is available at the end of the chapter

http://dx.doi.org/10.5772/65087

Abstract

Resilience is one of the three core properties of social-ecological systems, mixing adaptability and transformability. Flexibility can be defined in terms of diversity of procedures and the speed at which they can be mobilized by one organization. The analyses performed are presented in terms of levers that farmers can deploy to protect their management systems against market uncertainty. These levers differ depending on farmer standpoints, objectives, lessons learned, the collective organizations they work with, the standards and specifications they work to, etc. It is equally important to identify the interplays between overarching and underlying scale levels for the system studied and to hone in on the dynamics at work during periods of transition. Adaptive capacities of farm systems are closely linked to how the farmer perceives the situations to manage, according to his aims, to his behaviour face to risk and to his idea of what is his job. We propose to use different words to describe the properties of farming systems to cope with changes, according to the level within the system: "adaptive capacity" or "plasticity" for the animal level, "resilience" for the biotechnical level and "flexibility" for the whole system, including the manager. We think there is a real challenge working at each level on transition periods and processes, as farming systems will have more and more to adapt face to unpredic events.

Keywords: resilience, flexibility, adaptive capacities, uncertainty, crops systems, animal system

1. Introduction

Farm businesses, just like any other business enterprise, develop response strategies in order to cope with the many demands imposed on them and the uncertainties they face. The challenge for farmers lies in securing sustainability for their business, in a context where farming is subject to wide-reaching change and where farms are increasingly exposed to agronomic trends and climatic risks that the agricultural productivity model generally seeks to overcome by controlling processes and disengaging the effects of environmental disturbance.

Incorporating the precepts of sustainable development in order to build and assess new technical agricultural systems hinges on breaking away from the rationales underpinning these systems and moving towards more holistic objectives encompassing far more than the simple production output function [1]. There are two key drivers to this breakaway: (i) reinventing how researchers interact with the other actors involved in the process of developing new systems and their multiple outcomes [2, 3], and (ii) producing tools capable of quickly rendering a priori system assessments [4, 5] as a first step towards subsequently deploying the systems in compliance with complex multicriteria specifications [6]. This means that agronomists face the challenge of translating the impacts of integrating these dimensions into terms that farmers can understand and use to reshape their farm systems, taking into account new social and environmental factors [7, 8].

This reshaping redefines the farm business as a complex system that needs to be analysed not just in terms of its type but also the rationales driving how it operates [9, 10]. A few years ago, farming system researchers started using the notion of flexibility to define the capacity of a business to weather and adapt to economic uncertainty. The concept of resilience, as pioneered by Holling [11], has also been analysed in this setting, particularly when applied in more recent social-ecological systems [12]. "Flexibility" has been researched extensively in management science and industrial economics, whereas "resilience" has mainly been used in ecology (but also in social psychology; [13]). Our study will draw on illustrative examples to highlight how the notion of flexibility can prove useful for designing and assessing innovative technical systems.

2. Flexibility in management sciences

Industrial economics and management sciences understand the concept of flexibility [14, 15] as the capacity of a business or organization to re-adapt its structure and projects in response to environmental challenges (strategic flexibility) and to re-adapt its skillsbase, reorganize its workflows (workflow flexibility) and/or adjust its production methods as a response to unforeseen variations in *inputs* from outside (operational flexibility). The concept therefore appears relevant when analysing farmers' response strategies in the current climate governing agricultural production (characterized by regulatory developments, volatile agricultural prices, climatic variability, etc.). Tarondeau [15] (*ibid.*) stratified different sources of flexibility:

product flexibility (product range), process flexibility and inputs flexibility. The basic idea is that the capacity to cope with unknowns and carry the business forward is dependent on several factors, both material and non-material: the configurations of their technical production systems, their structures, their projects and their objectives [16]. Reix et al. [17] suggest that the drive for flexibility can be seen as the drive to maintain consistency in how the business is managed in response to a changing environment. Flexibility, as a system property, is not "given": it is built, shaped and "nurtured," and it has a cost [18]. Flexibility can be considered a competitive advantage insofar as it enables performance levels to be sustained in situations of uncertainty [19].

Different commentators use different terms as synonymous with or acceptances of the concept of flexibility, but there is a body of ideas that remain recurrent. Flexibility refers to organizational capacity [14, 15, 20–23, 28]. This means that the systems described are always management-led and that the organizational procedures governing their management constitute a source of flexibility for the system. In each case, flexibility is defined as an attribute that is inherent to humans, dependent on how they perceive situations to be addressed, their objectives, their level of risk aversion and the perception they hold of their business. Flexibility is a property that has to associate both change and stability, forming a paradox between permanence (continuity, mainstay) and change [16]. The authors see management flexibility as the result of constructive tension between what needs to be held onto and what needs changing. This same idea has been explored through analyses of how livestock farming systems work, with the notion of invariants [24]. The invariant acts as a backbone, a basis, a bottom line and the frame of reference for handling change (not everything has to change at the same time, otherwise the system risks getting disorganized or even collapsing into chaos). Flexibility is intrinsically dynamic. It can only be meaningfully studied in the long term, at multiperiod scale. Integrating flexibility into the analysis of a system or an organization presupposes that the decision-maker is looking to achieve short-term objectives while also securing a range of opportunities for the longer term [25]. In other words, a given decision may appear non-rational (or non-optimal) when analysed at timepoint t, but become entirely rational once events liable to arise at some point in the future are factored in (uncertainty preparedness). Indeed, the speed of response to these events is one of the key components of flexibility [15].

Furthermore, in every scenario, the concept of flexibility is also linked to the notion of interaction between the system/organization and its environment. It can therefore be measured and thus assessed, by quantifying the degree of control (according to the dual flexibility concept proposed by De Leeuw and Volberda [20]: controlled systems vs. independent systems) over environmental inputs (**Figure 1**).

The two paradigms coexist within a single system (controlling-controlled) and must therefore be analysed in tandem. However, the extent to which one paradigm dominates the other reveals specific system behaviours.

The organization as an environment-controlled system: in this configuration, the organization "copes with" environmental factors [16]. Flexibility hinges on accommodative processes [26], which hallmarks defensive behaviour in response to external perturbation [27]. The target objective for the system will be adaptation, stability, resilience to environmental forces and robustness. Systems unable to achieve this objective would be defined as vulnerable.

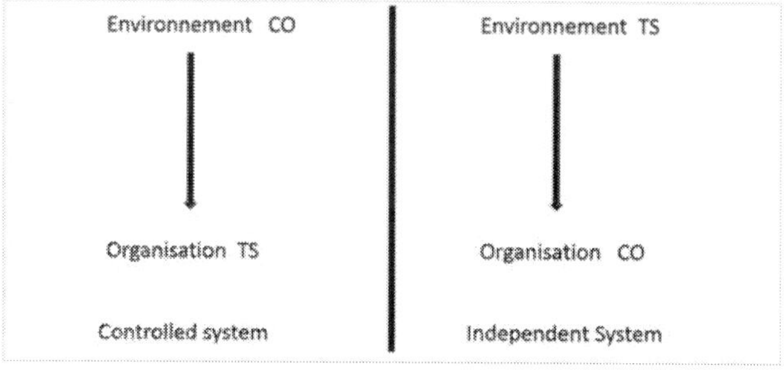

Figure 1. Organization of an environment-controlled system (left) and an environment-independent system (right) (concepts taken from Ref. [20]). TS = target system; CO = controlling organ. The arrows illustrate the direction of control exerted by the CO over the TS.

The organization as an environment-independent system: in this configuration, the organization seeks to subordinate all changes in its environment to the task of maintaining its objectives and its identity. Interactions with the environment are specified internally, and on a certain level, the environment is integrated into the organization. The processes deployed in the search for flexibility are assimilative processes, which hallmarks a pro-active pattern of behaviour that will respond to each perturbation by generating new behaviours, thereby expanding the range of adaptation options possible. These configurations define self-learning organizations with self-directed learning capacities.

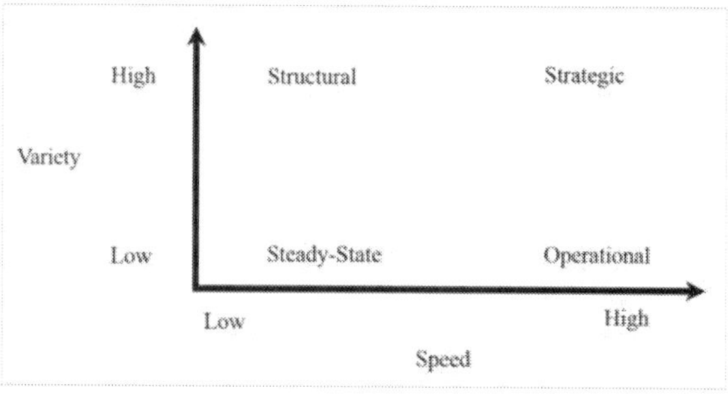

Figure 2. Different types of flexibility according to the number of planned procedures (vertical axis) and the speed at which they can be implemented (horizontal axis); adapted from Ref. [28].

De Leeuw and Volberda [20] encapsulated these two configurations by defining flexibility in terms of diversity of procedures and the speed at which they can be mobilized: (i) to increase the organization's environmental control capacities and (ii) to decrease the organization's environmental vulnerability. The authors define different types of flexibility according to the number of planned procedures and the speed at which they can be implemented (**Figure 2**).

3. Social-ecological resilience: a kind of flexibility?

The concept of resilience is borrowed from material physics as well as ecology as a means of describing the transformation and/or adaptive capacity of a material or ecosystem in response to stressors. In ecology, Holling [11] described resilience as the capacity of an ecological system or species to absorb challenges and then recover its initial configuration. The concept was then broadened to encompass shifts, learning and human-nature interactions [29]. Resilience was then extended to describe the mechanics of "anthropized" systems [30]. More recently, the concept of resilience has been applied to social-ecological systems, where humans are a governing actor [2, 12, 31–33]. The system is thus considered as a "learner," with a shift in the underlying idea from a return to the initial state following the perturbation towards a capacity to reconfigure itself while maintaining the core objectives and projects, where stakeholders can continue to plan for the future [2]. According to Ref. [34], there are three potential strategies capable of increasing the resilience of actively governed systems: increasing the system's buffer capacity (room for manoeuvre), scale-based governance (spatial and temporal scales) and creating opportunity for innovation (sources of change to system properties, learning capacity). These systems therefore have the ability to respond to perturbation by shifting into different stability domains rather than a single, "initial" steady state.

Walker et al. [35] outlined four main features of system resilience connected to the notions of steady state and initial state: (i) the amount of change that the system can tolerate without collapsing into an essentially different state, this idea works on the assumption that there is a threshold beyond which the system can no longer recover its initial configuration; (ii) the capacity to resist change, which is connected to properties like rigidity and robustness; (iii) vulnerability (precariousness), which is how close the system state is to the threshold cited under point 1; (iv) panarchy, which describes a system integrating a great many elements undergoing cross-scale interactions, and that the level of resilience depends on the different states and dynamics interplaying at the scales above and below.

Resilience can also be described in terms of successive system states over time. Holling [36] and Walker et al. [2] consider that ecological systems follow adaptive cycles comprising four successive phases. They posit that actively governed systems reproduce cyclic patterns of behaviour aligned to these four phases: a phase of accelerated growth (annotated r), followed by a longer phase of steady accumulation towards stability, associated with a progressive decline in resilience (K), then a sharp structural collapse (Ω) before another short phase of rebuilding and reorganization (α). Depending on the current phase of the system, a given disturbance (which can in fact be seen positively as the introduction of an accommodative stance) will not have the same effect.

4. Leverages to enhance flexibility in livestock systems

4.1. Different levers according to scale

Aaker and Mascarenhas [37] focusing on the means to enhance organizational flexibility outlined the following four levers centred on products, resources and management: (i) diversification of processes, business activities and products, running from broadening the range but also including activity in different marketplaces and extended use of different process technologies. In Ref. [38], the authors assert their notion of "relational flexibility" to account for the sources of adaptive capacity employed by livestock farmers through their marketing networks and the circuits they build or exploit to sell livestock; (ii) increasing inter-independence between production units; (iii) developing a base of potentially useful resources that are deployed not continually but on a case-by-case basis "should the need arise": functional redundancies, latent competencies, room for manoeuvre; (iv) minimizing workflow specialization, steering away from situations where tasks are accomplished by staff who have competencies deemed "necessary and sufficient" to complete the task. For example, Madelrieux et al. [39] clearly illustrate the flexibility achievable by a more collective workplace organization and workload breakdown in livestock farming systems.

Using two examples of farm systems (crop and livestock), we illustrate how these flexibility leverages can be deployed to minimize vulnerability to changes in the systems' environments. These two examples were chosen to demonstrate how the internal organization of the system (the sequencing of the system's structural components) and the system manager's perception of the environment act as complementary leverage for lending flexibility to farm production systems.

4.2. Animal contribution (plasticity) to system flexibility in an organic dairy system

The Mirecourt (INRA) research team prototypes sustainable dairy systems focused on agro-environmental sustainability. One system, tested since 2004, is a low-input grass-only system, in accordance with the specifications governing organic farming and based on the hypothesis that pasture-based systems are more sustainable [40].

This system is designed to introduce rulesets and animal and farmland management modes for achieving the objectives assigned to the system at the outset. In other words, the system aims to define how to achieve a result targeted at the outset without having to run through the conventional pattern of conducting experimental trials to measure results from different management condition sets established at the outset. Systems employing this strategy are designed to be sustainable in agro-environmental terms. More operationally, we posit that in order to cope with these objectives, the systems have to be self-sufficient (no importation of fertilizers or pesticides) and able to cope with unanticipated events, especially climatic events, since self-sufficiency can render systems more sensitive to natural variations in farmland properties.

The herd breed is split equally between two breeds (Holstein and Montbeliarde) in order to test the capacities of each breed to enable the system to achieve the objectives set. Maximizing grazed grass in the cow diet led to grouped calvings in late winter (February to April) in order to match the animals' energy requirements with grass availability. Under this management policy, cows produced 5132 kg milk/cow/year on average in 2005 and 2006: Holstein cows milked on average 400 kg milk/cow/year higher than Montbéliarde cows (respectively 5347 and 4947 kg milk/cow/year). However, at the end of the breeding period, 65% of dairy cows were pregnant in 2005 but only 27% at the corresponding timepoint in 2006. These very poor ratios affected herd sustainability, even though performance levels for replacement heifers were better (**Table 1**).

Year	2005			2006		
	Herd	Hn	Mo	*Herd*	Hn	Mo
Cows						
Success AI1 and AI2 (%)[1]	*54*	38	75	*23*	8	33
Fertility (%)[2]	*65*	52	81	*27*	17	33
Heifers						
Success AI1 and AI2 (%)	*79*	80	78	*71*	75	67
Fertility (%)	*86*	80	89	*88*	75	100

Hn: Holstein; Mo: Montbeliarde; AI: artificial insemination.

[1]Percentage of pregnant cows served once or twice.

[2]Percentage of animals calving after being served during the breeding period.

Table 1. Reproductive performances of dairy cows in 2005 and 2006, according to breed.

An analysis of individual animal management within the cow herd highlighted different groups. Each group corresponds to a specific calving date, which, in relation to turnout date, determines the feed diet at the beginning of lactation: a switch from winter feed to pasture grass.

The milk production of dairy cows calved after turnout increased very quickly (2–4 weeks) to maximum daily production, generating high energy requirements, which is detrimental to reproduction. The milk production of dairy cows calving at least one month before turnout showed a slower increase to maximum daily production (taking 8–12 weeks), with a smoother effect on energy balance and reproduction. Within these two configurations, Montbeliarde cows gave smoother lactation curves than Holstein cows (**Figure 3**). They were able to limit milk production, even when stimulated by turnout to grass, and thus gave better reproduction performances than Holstein cows.

In the grass-based systems, Montbeliarde cows offer more plasticity than Holstein cows. Secondly, shifting the calving period (January to March instead of February to April) should maximize the number of calvings before turnout to grass, thus lending the system more flexibility by enhancing reproductive performance.

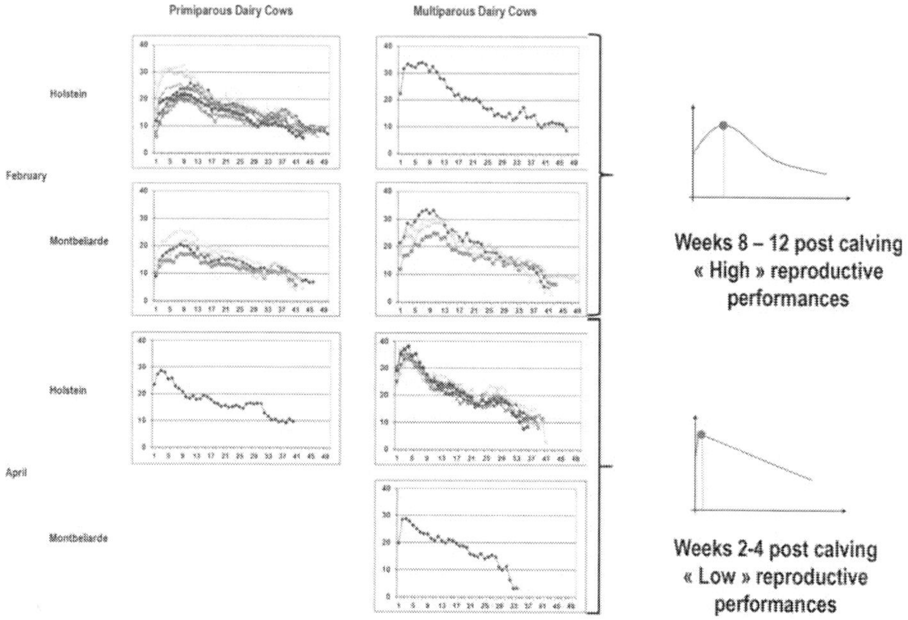

Figure 3. Individual lactation curves (milk yield in kg/cow/day throughout time after calving, in weeks) of Montbeliarde (Mo) and Holstein (Ho) dairy cows in 2005, according to the parity and to the calving period (February = at least 1 month before turnout vs. April = after turnout). On the right side, the average shape of curves for each period.

4.3. The collective workflows lever: flexibility in response to market uncertainty

The flexibility of suckler cattle farms is induced by commercial circuits: one of the features of suckler cattle farms is that they offer the possibility of selling livestock, and particularly females, at virtually any age. There are potentially over 15 different categories, with some breeders selling a minimal number of animal categories (n = 3: male calves, female calves and cows), whereas other systems offer a broader range comprising four or more different categories. Some systems always produce the same types of animal, whereas others gear themselves with options to change in response to climate events or market openings. There is also a heavy and practically range-independent variability in the number of buyers for the animals produced (**Figure 4**): a 2005 survey sampling livestock farmers ranged from one buyer for all animals up to seven different partners. Over and above buyer numbers, buyer status is also a critical criterion for livestock breeders. We have identified two different sets of strategic choices:

- Cooperatives vs. private buyers: some livestock farmers are convinced that cooperatives rob them of their freedom to market their products and thus refuse to help finance the running costs (premiums), in contrast to other farmers who strongly believe the cooperative represents their best interests, offering them a voice and a channel through which they can take action if problems arise. Finally, there is another category of livestock farmers who attach little importance to buyer status and who choose to sell their animals based on the prices they can get and how well they know and trust the buyer;

- Single buyer vs. several buyers: for farmers who work with a single buyer, the driving factor is the relationship of trust: the buyer understands how the farmer works and knows what animals are produced: negotiations are relatively straightforward, and sometimes a phone call is all that is needed. While the cattle farmer does need to make efforts to protect this special relationship (trust-system payments, sales spread across the year, etc.), in return they can expect the buyer to step in and make priority purchases when business is bad (security factor). In contrast, other farmers see the option of juggling between buyers as a way to take advantage of competition. If the market goes through a crisis, the farmer hopes to weather the storm by having a number of available buyers in order to sell their total livestock.

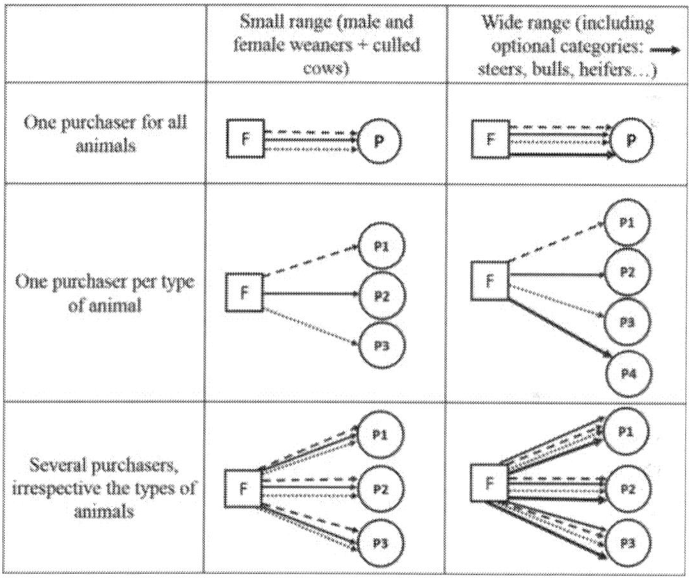

Figure 4. Different farmer (F) strategies for animal sales in livestock farming systems, combining range and number of purchasers (P); (one arrow corresponds to one specific category of animals sold, i.e. culled cows, weaned calves, heifers, bulls and steers).

The components of biophysical systems (plants, animals and soils; **Figure 5**) confer a relatively greater level of system-wide flexibility through their own, intrinsic properties: (i) delayed differentiation process: unicity, particularly for females from suckler cattle breeds, regardless of their end purpose and their age at sale [41]; (ii) plasticity, breed diversity and ability to adapt to different management strategies [42, 43]. Gaillard et al. [44] showed how Simmental breed diversity offered dairy farmers options to take up a more or less marked position on the intensified fodder system gradient, ranging from extensive 100% grassland systems to intensive corn silage-based systems.

Depending on the flexibility leverage deployed by the farmer [7], both the system components (structural dimensions) and their interplays (functional dimensions) will take on a certain measure of specificity. Furthermore, this distinction picks up on the distinction made by Alcaras and Lacroux [16] between the stability of an organization's structure and the stability of an organization's target objectives: (i) the "size" lever: reproductive capacities, useful lifespan and carcass yield, for animals that farmers can no longer select to work with once they opt to increase the size of their holding through internal growth (zero buy-in); (ii) the "responsiveness" lever (short-range opportunity-taking): versatility, ability to handle change (feed type and volume), malleability, breed mix, capacities for out-of-season production; (iii) the "collective workflows/technicity" lever: quantitative performance, standardized high-tech information system, records; (iv) the "room for manoeuvre" lever: versatility, simplicity, hardiness.

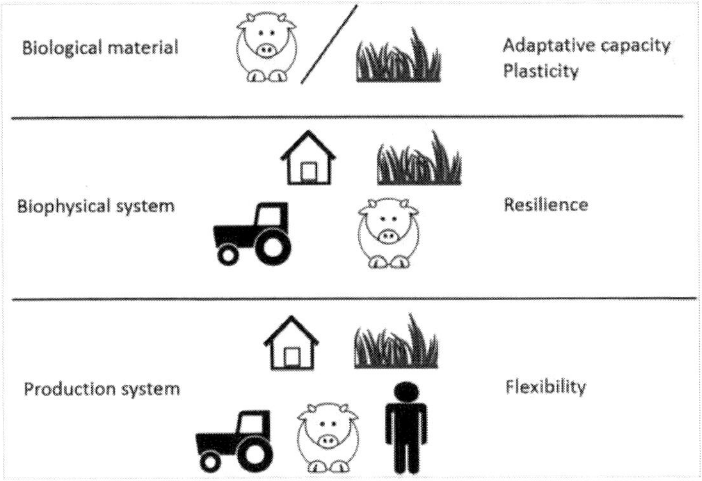

Figure 5. Descriptors assigned to adaptive capacities according to level of organization in the functional analysis of production systems.

5. Discussion and conclusion

The foundations of resilience analysis have progressively shifted towards the foundations of flexibility analysis. Our assertion is based on qualifying the set of properties that will enable a system to secure sustainability by restricting the use of the two terms to different levels of organization (**Figure 5**): "flexibility" to cover the level overarching the entire production system and "resilience" to cover the underlying level of the biophysical (or operant learning) system. The terms used at the next level down, comprising the organic system entities such as plants and animals, would be "plasticity" and "adaptive capacity" as employed in Ref. [42]. The three examples of production systems highlighted earlier share a common denominator in that they are all "extensive" systems, that is, where productivity per surface unit of land is not maximized compared to intensive systems. A clear pattern emerged, wherein the adaptive capacities of these systems are perceived differently under the two scenarios. The design and development of intensive systems (high production per surface unit of land) consisted then, as now, in targeting measures capable of absorbing the negative effects of increasing performance. This means that for the animals, the primary property needed is "robustness," that is, the ability to produce a lot and regularly, regardless of the environmental disturbances.

The levers that farmers can deploy to protect their management systems against market uncertainty will differ depending on farmer standpoints, objectives, lessons learned, the collective organizations they work with, the standards and specifications they work to, etc. Therefore, in order to properly analyse the attributes of systems that make them less vulnerable to unknowns, the focus should be directed towards the information systems employed by farm system managers [45]. It is equally important to identify the interplays between overarching and underlying scale levels for the system studied (panarchy) and to hone in on the dynamics at work during periods of transition.

Literature review combined with the examples compiled reveals that studies directed at developments and changes in farm systems harnessing ecological-biological (animals, plants, etc.) and human-social (farmers' strategies and objectives) dimensions can use the notion of flexibility to gain a sharper and more explicit analysis of the interactions between these dimensions.

The move to revitalize the analytical framework governing livestock farming systems has to explicitly factor in dimensions stemming from interactions between animal production science and social sciences (formalization of livestock farmer strategies, workflow organization; [46]) as well as between ecology (resilience) and management science (flexibility). The target is to combine the analytical perspectives on (i) the regulatory properties of management-led biological systems (such as the herd, whose dynamics are shaped by interactions between human decisions and the biological functions of the animals; [43, 47]) and the leverages capable of parrying the effects of climatic risks and economic unknowns (types of product, relations with downstream factors, socio-technical networks).

There has been a key turning point in the way agronomics researchers have addressed the issue of performance in farm production systems. There has been a move away from focusing on ways to control or increase quantitative performance metrics (although there are shades of ecological intensification policy that still encourage this kind of outlook; [48]) and towards other rationales, such as "multicriteria" system design and assessment frameworks. Looking at the issues left unresolved and the various standpoints on offer, we have identified at least two courses of action:

- The interplay, or rather the fitting of abilities between production system components (system entities) and the type of system environment. This standpoint leads to a subsequent issue of whether there are advantages to be drawn from preserving certain specific animal or plant genotype characteristics that are underrepresented or tend to pale in comparison when balanced against the yield capacities of different breeds and the so-called improved crop varieties.
- The advantages of mixed farm systems combining different animal breeds/plant species, where the farmer is hedging on complementarity between the properties of each breed/species to cope with climatic unknowns (species offering different hardiness or which develop at different periods of the year) or variations in market prices (which have different effects on different farm outputs).

Approaches based on concepts and theories borrowed from disciplines such as ecology and management science are particularly fruitful for fuelling reflective thinking and reframing analyses in agronomics science when the aim is to investigate the dynamics of change and the adaptability of farm in response to situations of uncertainty.

For farmers, the art of farm management resides in tackling head-on how they define and readjust the production objectives set, how they lead negotiations with other farm stakeholders in order to achieve these objectives given the resources available, how they tackle uncertainty and how they tackle opportunity. These are all complex adaptation processes occurring at the interface between the farm and its environment, which emerge not only in the decisions taken but also in the short-term and long-term practices that we have termed "flexibility." Our analysis of these processes applied to three real-world systems enabled us to highlight a handful of principles governing farm business flexibility. First, the situational contextualization: flexibility is dependent not only on the technical features of the production system components (plasticity) but also on the socio-economic environment in which the businesses evolve; second comes the collectiveness component: flexibility becomes greater as the business integrates the collective dimension of farm activity, even if the overriding aim is to maintain decision-making autonomy over the production system. Finally, from the methodology standpoint, trials led at our experimental farm station have prompted us to continue investigations into methods for qualifying and if possible even quantifying the sustainability of farm structures in interaction with their environment, factoring in the different farm-structuring organizational levels. This research will ultimately be used for inter-farm comparisons integrating on-farm production system adaptability over time.

Acknowledgements

This work was carried out with the financial support of the "ANR—Agence Nationale de la Recherche—The French National Research Agency" under the "Programme Agriculture et Développement Durable," project "ANR-05-PADD-04-01, Discotech."

http://dx.doi.org/10.5772/65087

Author details

Stéphane Ingrand[1,2,3,4*], Laura Astigarraga[5], Eduardo Chia[6], Xavier Coquil[7], Christophe David[8] and Jean-Louis Fiorelli[7]

*Address all correspondence to: stephane.ingrand@clermont.inra.fr

1 Inra, UMR, Métafort, Méthamis, France

2 Agro Paris Tech, UMR, Aubière, France

3 Irstea, UMR, Aubière, France

4 Clermont Université, VetAgro Sup, UMR, Clermont-Ferrand, France

5 Universidad de la Republica, Facultad de Agronomia, Montevideo, Uruguay

6 UMR Innovation, Montpellier Cedex, France

7 Inra UR Aster, Mirecourt, France

8 Isara, Lyon Cedex, France

References

[1] Sabatier R, Durant D, Hazard L, Lauvie A, Lecrivain E, Magda D, Martel G, Roche B, de Sainte Marie C, Teillard F, Tichit M. Towards biodiversity-based livestock systems: review of evidence and options for improvement. CAB Reviews. 2015;10(20):1–13. doi: 10.1079/PAVSNNR201510025

[2] Walker B, Carpenter S, Anderies J, Abel N, Cumming G, Jansen M, Lebel L, Norberg J, Peterson GD, Pritchard R. Resilience management in socio-ecological systems: a working hypothesis for a participatory approach. Conservation Ecology and Society. 2002;6(1):14. Available from: http://www.consecol.org/vol6/iss1/art14 [accessed 2016-07-25]

[3] Godart O, Hubert B. Le développement durable et la recherche scientifique à l'Inra: Inra Paris; 2002. 58 p. Available from: http://www.youscribe.com/catalogue/presentations/savoirs/techniques/le-developpement-durable-et-la-recherche-scientifique-a-l-inra-395481 [accessed 2016-07-25]

[4] Sadok W, Angevin F, Bergez JE, Bockstaller C, Colomb B, Guichard L, Reau R, Doré T. Ex ante assessment of the sustainability of alternative cropping systems: implications

for using multicriteria decision-aid methods. A review. Agronomic Sustainable Development. 2008;28:163–174. doi:10.1051/agro:2007043

[5] Gouttenoire L, Cournut S, Ingrand S. Farm-scale models built with farmers converting and converted to organic farming. Agronomy for Sustainable Development. 2012;33(3): 609–619. doi:10.1007/s13593-012-0125-6

[6] Coquil X, Fiorelli JL, Mignolet C, Blouet A, Foissy D, Trommenschlager JM, Bazard C, Gaujour E, Gouttenoire L, Schrack D. Assessing the agro-environmental sustainability of organic mixed-crop dairy systems on the basis of a multivariate approach. Innovations Agronomiques. 2009;4:239–247. Available from: https://www.researchgate.net/publication/277800762_Assessing_the_agro-environmental_sustainability_of_organic_mixed-crop_dairy_systems_on_the_basis_of_a_multivariate_approach [accessed 2016-07-25]

[7] Lémery B, Ingrand S, Dedieu, B, Degrange B. Agir en situation d'incertitude: le cas des éleveurs de bovins allaitants. Economie Rurale. 2005;288:57–69. doi:10.4000/economierurale.2718

[8] Rigolot C, Roturier S, Dedieu B, Ingrand S. Climate variability drives livestock farmers to modify their use of collective summer mountain pastures. Agronomy for Sustainable Development. 2014;34:899–907. doi:10.1007/s13593-014-0224-7

[9] Gafsi M. Exploitation agricole et agriculture durable. Cahiers Agricultures. 2006;15(6): 491–497. doi:10.1684/agr.2006.0035

[10] Chia E, Dugué P, Sakho-Jimbira S. Can family farms be considered as institutions? Cahiers Agricultures. 2006;15(6):498–505. doi:10.1684/agr.2006.0027

[11] Holling CS. Resilience and stability of ecological systems. Annual Review of Ecological Systems.1973;4:1–23. doi:10.1146/annurev.es.04.110173.000245

[12] Anderies JM, Jansen, M.A, Ostrom E. A framework to analyze the robustness of social-ecological systems from an institutional perspective. Ecology and Society. 2004;9(1):18. Available from: http://www.ecologyandsociety.org/vol9/iss1/art18/[accessed 2016-07-25]

[13] Tisseron S. La résilience. Coll. Que sais-je? Edition PUF; 2007. 128 p. Available from: http://www.cairn.info/la-resilience--9782130577959.html [accessed 2016-07-25]

[14] Cohendet P, Llerena P, editors. Flexibilités, risque et incertitude dans la théorie de la firme. In Flexibilité, information et décision, Paris, Éditions Économica;1989

[15] Tarondeau JC. Approches et formes de flexibilité. Revue française de gestion, mars-avril-mai.1999;66–71.

[16] Alcaras JR, Lacroux F. Planifier, c'est s'adapter. Economies et sociétés Sciences de gestion, Série S.G. 1999;26–27:7–37.

[17] Reix R. « Flexibilité », Simon, Y, Joffre P. (sous la direction de), Encyclopédie de Gestion, Tome II, Paris, Economica; 1997. p. 1407–1420.

[18] Chia E, Marchesnay M. Un regard des sciences de gestion sur la flexibilité: enjeux et perspectives. In: Dedieu B, Chia E, Leclerc B, Moulin C.H, Tichit M, (Eds), "L'élevage en mouvement: flexibilité et adaptations des exploitations d'herbivores". Editions Quae; 2008. p. 23–36. Available from: http://www.quae.com/fr/r1308-l-elevage-en-mouvement.html [accessed 2016-07-25]

[19] Dreyer B, Gronhaug K. Uncertainty, flexibility and sustained competitive advantage. Journal of Business Research. 2004;57(5): 484–494. doi:10.1016/s0148-2963(02)00315-6

[20] De Leeuw ACJ, Volberda HM. On the concept of flexibility: a dual control perspective. Omega.1996;24(2):121–139.doi:10.1016/0305-0483(95)00054-2

[21] Evans JS. Strategic flexibility for high technology manoeuvres: a conceptual framework. Journal of Management Studies. 1991;28(1): 69–89.doi:10.1111/j.1467-6486.1991.tb00271.x

[22] Sanchez R. Strategic flexibility in product competition. Strategic Management Journal. 1995;16(S1):135–159. doi:10.1002/smj.4250160921

[23] Weick KE. Psicología social del proceso de organización. Colombia: Fondo Educativo Interamericano. 1982

[24] Moulin CH, Ingrand S, Lasseur J, Madelrieux S, Napoléone M, Pluvinage J, Thénard V. Comprendre et analyser les changements d'organisation et de conduite de l'élevage dans un ensemble d'exploitations: propositions méthodologiques. In: Dedieu B, Chia E, Leclerc B, Moulin CH, Tichit M, editors. L'élevage en mouvement: flexibilité et adaptations des exploitations d'herbivores. Editions Quae; 2008. p. 181–196.

[25] Lev L, Campbell DJ. The temporal dimension in farming systems research: the importance of maintaining flexibility under conditions of uncertainty. Journal of Rural Studies. 1987;3(2): 123–132. doi:10.1016/0743-0167(87)90028-3

[26] Piaget J. La naissance de l'intelligence chez l'enfant, Delachaux et Niestlé. 1977.

[27] Le Moigne JL. La modélisation des systèmes complexes, Dunod; 1990.

[28] Volberda H. Towards the flexible form: how to remain vital in hypercompetitive environments. Organisation Science. 1996;7(4):359–374. doi:10.1287/orsc.7.4.359

[29] Gunderson LH, Holling CS, Light SS editors. Barriers and bridges to the renewal of ecosystems and institutions. Columbia University Press. New York; 1995.

[30] Ludwig D, Walker B, Holling CS. Sustainability, stability and resilience. Conservation Ecology. 1997;1(1):7. Available from: http://www.ecologyandsociety.org/vol1/iss1/art7/ [accessed 2016-07-25]

[31] Carpenter S, Walker B, Anderies JM, Abel N. From metaphor to measurements: resilience of what to what? Ecosystems. 2001;4(8):765–781.doi:10.1007/s10021-001-0045-9

[32] Milestad R, Darnhofer I. Building Farm Resilience: the prospects and challenges of organic farming. Journal of Sustainable Agriculture. 2003;22(3):81–97.doi:10.1300/j064v22n03_09

[33] Folke C. Resilience: the emergence of a perspective for social-ecological systems analysis. Global Environmental Change. 2006;16: 53–267. doi:10.1016/j.gloenvcha.2006.04.002

[34] Gunderson LH. Ecological resilience in theory and application. Annual Review of Ecological Systems. 2000;31(1):425–439. doi:10.1146/annurev.ecolsys.31.1.425

[35] Walker B, Holling CS, Carpenter SR, Kinzig A. Resilience, adaptability and transformability in socio-ecological systems. Ecology and Society. 2004;9(2):5. Available from: http://www.ecologyandsociety.org/vol9/iss2/art5/[accessed 2016-07-25]

[36] Holling CS. The resilience of terrestrial ecosystems: local surprise and global change. In: sustainable development of the biosphere. Clark WC, Mumn ME, editors. Cambridge University Press; 1986. p. 292–317.

[37] Aaker DA, Mascarenhas B. The need for strategic flexibility. Journal of Business Strategy. 1984;5(2):74–82. doi:10.1108/eb039060

[38] Dedieu B, Chia E, Leclerc E, Moulin CH, Tichit M. editors. L'élevage en mouvement. Flexibilité et adaptation des exploitations d'herbivores. Editions Quae. Update Sciences & Technologies; 2008. 294 p.

[39] Madelrieux S, Dedieu B, Dobremez L, Girard N. Patterns of work organisation in livestock farms: the ATELAGE approach. Livestock Science. 2008;121(1):28–37. doi:10.1016/j.livsci.2008.05.014

[40] Kaine GW, Tozer PR. Stability, resilience and sustainability in pasture-based grazing systems. Agricultural Systems. 2005;83(1):27–48. doi:10.1016/j.agsy.2004.03.001

[41] Astigarraga L, Ingrand S. Production flexibility in extensive beef farming systems. Ecology and Society. 2011;16(1):7. Available from: http://www.ecologyandsociety.org/vol16/iss1/art7/[accessed 2016-07-25]

[42] Blanc F, Bocquier F, Agabriel J, D'Hour P, Chilliard Y. La composante animale de la flexibilité des systèmes d'élevage. In: Dedieu B, Chia E, Leclerc B, Moulin CH, Tichit M, editors. L'élevage en mouvement: flexibilité et adaptations des exploitations d'herbivores. Editions Quae; 2008. p. 73–94.

[43] Ollion E, Ingrand S, Delaby L, Trommenschlager JM, Colette-Leurent S, Blanc F. Assessing the diversity of trade-offs between life functions in early lactation dairy cows. Livestock Science. 2016;183:98–107.doi:10.1016/j.livsci.2015.11.016

[44] Gaillard C, Ingrand S, Blanquet J, Devevey C, Dumont R. Relations entre pratiques de finition et performances à l'abattage de vaches de réforme Simmental Française: intérêt du caractère mixte de la race. Inra Productions Animales. 2007;20:327–336. Available from: http://www6.inra.fr/productions-animales/2007-Volume-20/Numero-4-2007/Relations-entre-pratiques-de-finition-et-performances [accessed 2016-07-25]

[45] Magne MA. Modéliser le système d'information des agriculteurs. Le cas des éleveurs de bovins allaitants [thesis]. Université Montpellier II; 2007. Available from: http://metafort.irstea.fr/wp-content/uploads/2015/11/MAMagne.pdf [accessed 2016-07-25]

[46] Vavra M. Sustainability of animal production systems: an ecological perspective. Journal of Animal Science. 1996;74:1418–1423. Available from: http://oregonstate.edu/dept/eoarcunion/sites/default/files/publications/VJAS_74_1418.pdf [accessed 2016-07-25]

[47] Tichit M, Ingrand S, Moulin CH, Cournut S, Lasseur J, Dedieu B. Analyser la diversité des trajectoires productives des femelles reproductrices: intérêts pour modéliser le fonctionnement du troupeau en élevage allaitant. Inra Productions Animales. 2004;17(2):123–132. Available from: http://www6.inra.fr/productions-animales/2004-Volume-17/Numero-2-2004/Analyser-la-diversite-des-trajectoires-productives-des-femelles-reproductrices [accessed 2016-07-25]

[48] Griffon M. Pour des agricultures écologiquement intensives. Des territoires à haute valeur environnementale et de nouvelles politiques agricoles. Les leçons inaugurales du groupe ESA;2007. 73 p. Available from: http://www.agrobiosciences.org/article.php3?id_article=2276 [accessed 2016-07-25]

Chapter 2

Do Organic Livestock Farms Differ from Low-Input Conventional Ones? Insights Based on Beef Cattle in Southern Europe

Alfredo J. Escribano, Paula Gaspar,
Francisco J. Mesías and Miguel Escribano

Additional information is available at the end of the chapter

http://dx.doi.org/10.5772/65094

Abstract

The objective of this study is to determine whether there are clear differences between conventional (but low-input) and organic beef cattle farms located in the Southwest of Spain. Thirty-three conventional and 30 organic farms were compared in terms of structure, technical management, and performance. The results showed that organic farms ('All Organic') mainly focus on the production of calves at weaning age, which are fattened in conventional holdings ('Organic 1'; $n = 22$). The remaining organic farms ('Organic 2'; $n = 11$) showed to participate in almost all stages of the agri-value chain. 'Conventional' farms were mainly dedicated to producing calves at weaning age (similarly to Organic 1). Organic 1 had the smallest herd size (80.18 livestock units (LU), $p < 0.05$). Organic 2 showed greater presence of indigenous breeds (62.08%, $p < 0.05$). Conventional farms proved to bear higher feed and veterinary costs per area (161.59 and 17.87 €/ha; $p < 0.01$ and $p < 0.05$, respectively), but Organic 2 had higher feed costs per LU. Therefore, Conventional and All Organic were quite similar, and differences depended mainly on farm structure. Hence, being either conventional or organic does not seem to be a valid criterion for drawing conclusions regarding the benefits or characteristics of each system.

Keywords: semiarid, production systems, sustainability, sustainable agriculture, mediterranean, drylands

INTECH
open science | open minds

1. Introduction

Organic livestock farm numbers have sharply increased in recent years [1] as an adaptive response for increasing farm profitability (through subsidies and price premiums). However, before implementing any production systems, an analysis of the similarities and differences between both the current and the potential new farm configurations should be carried out, since it will provide a wider view of the chances of success after the change.

For this purpose, the methodological process of farm characterisation is essential as it allows in-depth understanding of the operation of livestock production systems, which is key to improving their management, economic performance and overall sustainability. Thus, Rodríguez et al. [2] stated that farm viability relies on specific management practices that are suitable for the specific socioeconomic and environmental context of the farm, and this should be based on the knowledge of the characteristics and performance of the production systems.

Subsequently, several researchers have conducted studies for characterising farms according to various parameters. Some authors have focused on livestock species reared on the farms. Thus, different authors [3–10] have studied and characterised beef cattle farms by means of descriptive and/or cluster analysis on the basis of technical, structural, economic and/or social indicator. However, to our knowledge, there are no available studies that comparatively characterise organic and pasture-based or low-input conventional beef cattle farms, contextualizing such analysis within the evolution of the production systems under study. We therefore believe that this is a particularly appropriate time to conduct the present study. This would be of even more interest if the farms studied were located in complex agro-ecosystems with great value and externalities from the socio-economic and environmental points of view.

The present study was carried out with the following aims: (i) to shed light on the gap of knowledge existing due to the lack of studies that compare the characteristics of conventional and organic beef cattle farms and (ii) to find similarities and differences between organic and low-input conventional beef cattle farms. For this purpose, a characterisation (technical management, structure and economic performance) of the farms located in the '*dehesa*' was carried out.

2. Material and methods

2.1. Study area

The study area was the *dehesa* located in the region of Extremadura (Southwest of Spain). From a climatic point of view, it enjoys annual average temperatures of 16–17°C, with mild winters (average temperature of 7.5°C) and hot and dry summers (the average mean temperature is greater than 26°C, exceeding 40°C in the hottest months, which correspond to a Mediterranean continental semiarid climate. Its rainfall pattern is irregular (300–800 mm). Soils are shallow, acidic and of low fertility. Due to these characteristics, the availability of grazing resources is reduced and unstable [11–13].

2.2. Sample selection

A sample of farms in the beef cattle sector located in the *dehesa* of Extremadura was selected. Due to a lack of official statistics on figures and locations of *dehesa* farms in Extremadura, the sampling was non-probabilistic by quotas. The number of farms surveyed was 63, in line with other studies analysing livestock production systems [4, 14–17]. Apart from the number of farms, various criteria (already explained by Escribano et al. [17, 18]) were used to select the farms with the aim of obtaining an indicative sample of the various beef cattle production systems located in *dehesas*. The criteria used are summarised below:

- Predominant species and productive orientation: beef cattle.

- Herd size: over 25 adult cows, in order to differentiate between small and commercial farms.

- Geographical and forest-related aspects: the study includes farms located in different areas within the *dehesa* (geographical criteria) with different woodland densities (forest criteria similar to that followed in previous studies in the area [14]). **Figure 1** shows farms' spatial distribution and type of *dehesa* in which they were based on.

- Organic farms: all organic farms had already finished their conversion period to the organic system.

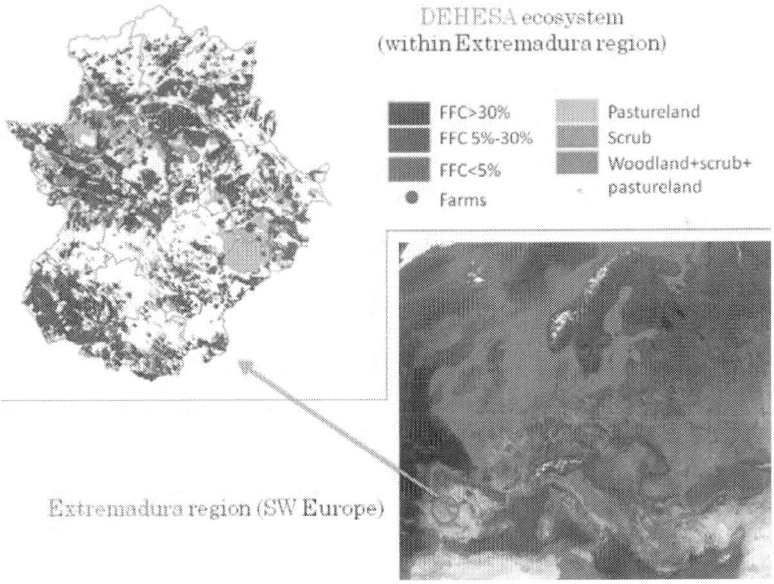

Figure 1. Dehesa location and different land cover characteristics. Forest fractional cover (FFC): Fraction of the land covered by the vertical projection of the tops of trees.

Finally, 63 farms (30 Conventional and 33 Organic farms were selected, thus achieving: a sample size similar to that of other studies characterising livestock farms [4, 5, 10, 14]; similar sample sizes of organic and conventional farms, thus allowing an adequate comparative analysis of both sectors.

2.3. Selection of parameters

In order to select the most appropriate indicators to analyse the farms under study, two main steps were followed. Firstly, the scientific literature addressing the structural and technical-economic points of view was reviewed. The selection of consistent and similar indicators allowed carrying out comparisons with studies on the topic. Moreover, economic parameters were created following the economic accounts for agriculture in the community [19] and the adaptation to dehesa livestock farms already carried out in previous studies [14, 15, 20]. As a consequence, the discussion of the results was consistent and the achievement of the aims of the present study was possible.

Finally, the selected set of indicators were confirmed to be in agreement with the recommendations of Lebacq et al. [21]: relevance, representativeness of the system, measurable, value to the end user, no ambiguity, no redundancy, and predictive.

2.4. Data collection

Data were collected from farms by means of a questionnaire in the year 2010. The questionnaire was developed according to selected indicators. These included information on structure (farms and herd characteristics: sizes, infrastructure, etc.), technical management, production results, economic data and social aspects. Subsequently, data were collected by the first author directly at the farms, followed by structured and semiclosed interviews with farm managers. Farmers' answers were the sources of information for all indicators. All these processes were carried out in accordance with the methodology used by several authors who analysed similar aspects of livestock farms [2, 6, 8, 10, 14, 15, 17, 22–27].

2.5. Analysis

The statistical analyses included descriptive statistics for the full sample of farms. Subsequently, an ANOVA test was applied to all parameters, as all of them are quantitative ones. This allowed comparing all farms following two approaches. First, conventional farms were compared to organic farms in order to compare the two production systems as a whole (Conventional vs. All Organic). Secondly, farms were compared based on three classifications that are explained in the next section: (i) Conventional farms; (ii) Organic 1 farms; (iii) Organic 2 farms. This approach offered insight into each of them, so that more valuable and precise conclusions about the organic beef cattle sector could be made. Statistical analyses were performed using SPSS v. 20.

3. Results

3.1. Farm types

After collecting data and creating the database, it was noted that, based on the aspects studied, organic farms could clearly be subdivided into two production systems, so it was decided that a classification of the farms selected needed to be made, with the resulting following groups:

- Conventional; found as "Conv." in the tables (n = 30): This grouped conventional farms. With regards to the situation of the beef cattle sector in the *dehesa*, these farms were mostly focused on calf rearing (calf fattening was almost nonexistent, so these farms mainly sold their calves at weaning age (5–6 months old and 160–220 kg of live weight; see **Table 1**).

Parameters	Conv. (n = 30)	Org. 1 (n = 22)	Org. 2 (n = 11)	Sig. 1[4]	Sample (n = 63)	SD	All Organic (n = 33)	Sig. 2[5]
UAA[1]	275.80	223.72	337.84	0.378	268.44	223.34	261.76	0.806
Owned area/UAA	0.64	0.54	0.55	0.541	0.59	0.44	0.55	0.390
Wooded land/UAA	0.46	0.47	0.77	0.101	0.52	0.43	0.57	0.336
Crop area/UAA	0.00	0.00	0.00	0.576	0.00	0.01	0.00	0.334
Bovine LU[2]	104.92^{ab}	74.33^{a}	124.83^{b}	0.016*	97.72	52.14	91.16	0.299
Ovine LU	6.78	5.36	15.37	0.496	7.78	30.38	8.69	0.805
Swine LU	0.00	0.50	0.77	0.445	0.31	1.55	0.58	0.138
Total LU	111.70^{ab}	80.18^{a}	140.95^{b}	0.024*	105.80	63.33	100.44	0.485
Bovine LU/Total LU	0.98	0.96	0.92	0.452	0.96	0.13	0.85	0.369
Total stocking rate[3]	0.73	0.50	0.44	0.312	0.60	0.64	0.48	0.131

a, b, c Mean values with different letters in the same row are significantly different. * $p<0.05$, ** $p<0.01$, *** $p<0.001$. SD: standard deviation. [1]UAA: Utilized Agricultural Area. [2]LU: Livestock Units. 1 cow = 1 LU; 1 sheep = 0.12 LU; 1 sow = 0.37 LU. [3]Total Stocking rate = LU/ha UAA. [4]Analysis of Variance of the groups Conventional, Organic 1 and Organic 2. [5]Analysis of Variance of the groups Conventional vs. All Organic.

Table 1. Herd and farm structure. Mean values, standard deviation and level of significance.

- Organic 1; found as "Org. 1" in the tables (n = 22): These farms were producing under the organic system, but none of them fattened their calves. On the contrary, they were marketed with almost the same characteristics (age and live weight) and price as the conventional ones (check **Table 1** to observe the similarity with conventional farms).

- Organic 2; found as "Org. 2" in the tables (n = 11): Organic farms that fattened and sold their calves under the organic system.

Furthermore, organic farms were also analysed as a whole in a group called "All Organic" (n = 33).

3.2. Farm structure and management

3.2.1. *Farm and herd characteristics*

The average size of farms was 268.44 ha utilized agricultural area (UAA) (**Table 2**). Organic 2 farms were larger than 'All Organic', but the high variability within the sample did not allow identifying significant differences between this group and Organic 1. With regard to herd size, All Organic farms were also similar to Conventional farms, and an important variation within farms was identified in relation to the mean cattle herd size (140.95 livestock units (LU) in Organic 2 vs. 80.18 in Organic 1, $p < 0.05$).

Parameters	Conv. (n = 30)	Org. 1 (n = 22)	Org. 2 (n = 11)	Sig. 1[6]	Sample (n = 63)	SD	All Organic (n = 33)	Sig. 2[7]
Replacement rate (%)[1]	11.98	11.92	13.29	0.922	12.19	1.23	12.38	0.875
Cows/bull (N°)	31.01	30.67	28.29	0.844	30.42	1.68	29.88	0.740
Estrous synchronisation (%)[2]	6.70	0.00	0.00	0.321	3.20	–	0.00	0.132
Artificial insemination (%)[3]	6.70	4.50	0.00	0.592	4.80	–	3.00	0.658
Length of mating period (months)	10.40	10.70	10.14	0.922	10.46	0.29	10.52	0.846
Fertility rate (%)[4]	85.15	77.70	81.49	0.187	81.91	1.82	78.97	0.091
Age at first calving (month)	30.68	33.45	33.68	0.197	32.17	0.79	33.53	0.074
Calving interval (days)[5]	346.50	33500	343.64	0.165	341.98	2.74	337.88	0.117
Calves born/cow/year (N°)	0.85	0.78	0.81	0.187	0.82	0.02	0.78	0.091
Weaned calves/cow/year (N°)	0.81	0.71	0.65	0.061	0.75	0.03	0.69	0.025*
Age at weaning (months)	5.86	5.82	6.00	0.886	5.87	0.13	5.88	0.944
Live weight at calving (kg)	202.33^{a}	190.91^{b}	193.18^{ab}	0.037*	196.75	2.12	191.67	0.011*
Calves sold at weaning age/cow/year (N°)	0.63^{a}	0.66^{a}	0.27^{b}	0.000***	0.58	0.03	0.53	0.000***
Fattened calves sold/cow/year (N°)	0.07^{a}	0.00^{a}	0.45^{b}	0.000***	0.11	0.03	0.15	0.000***
Fattened calves/total calves sold	0.09^{a}	0.00^{a}	0.64^{b}	0.000***	0.15	0.30	0.21	0.119

a, b, c Mean values with different letters in the same row are significantly different. * p<0.05, ** p<0.01, *** p<0.001. [1]Calculated as the annual average proportion of heifers bred for reproduction/number adult. [2]Annual average proportion of cows synchronized/total adult cows in the farm. [3]Average annual proportion of inseminated cows/total adult serviced cows in the farm. [4]Annual average proportion of: live births/serviced cows. [5]Annual average number of days from calving to calving in the adult cows of the farms. [6]Analysis of Variance of the groups Conventional, Organic 1 and Organic 2. [7]Analysis of Variance of the groups Conventional vs. All Organic.

Table 2. Reproductive management and performance, and productive orientation. Mean values, standard deviation and level of significance.

In relation to the various land uses and the type of ownership, it was seen that 59% of land was in property (owned area/UAA in percentages). Fifty-two percent of UAAs had tree presence. Moreover, crop areas were almost inexistent.

3.2.2. *Reproductive management and performance, and production results*

Estrous synchronisation was only carried out in the 3.20% of farms held. This practice was only observed in conventional farms, since it is not permitted in organic farming. Accordingly, only 4.80% of farms opted for artificial insemination, with all of them also carrying out natural mating, such that the use of either one or another technique was not exclusive. This scarce use of these reproductive techniques is typical in low-input beef cattle farms.

Calves weaned in All Organic had lower weights than those belonging to the conventional group, and Organic 2 farms sold less weaned calves per cow in total, thus showing a lower productivity in this regard. However, Organic 2 and All Organic sold more fattened calves per cow and also sold a higher proportion of fattened calves/total calves sold. These differences were due to the fact that the Organic 2 group was composed entirely of fattening farms, while all Organic 1 farms solely marketed calves at the age of weaning. Similarly to the Organic 1 group, 83.33% of the farms belonging to the conventional group did not carry out the fattening of any of the calves that they produced. These facts about the composition of the groups also influenced the differences between these indicators for yearlings sold per cow and calf weight at weaning.

3.2.3. *Breeds*

The breed distribution of organic farms is also an important issue, as autochthonous breeds are preferable for this production model, as indicated by Regulation 834/2007 [28]. **Table 3** shows the composition by breed of the farms.

Parameters	Conv. (n = 30)	Org. 1 (n = 22)	Org. 2 (n = 11)	Sig. 1[1]	Sample (n = 63)	SD	All Organic (n = 33)	Sig. 2[2]
Purebred autochthonous cows (%)	20.11^{a}	30.76^{ab}	62.08^{b}	0.015*	41.83	31.16	41.20	0.045*
Purebred foreign cows (%)	8.86	11.25	6.74	0.854	22.31	9.33	9.75	0.877
Purebred cows (%)	28.97^{a}	42.01^{ab}	68.82^{b}	0.027*	42.85	40.48	50.95	0.041*
Purebred autochthonous bulls (%)	13.02	15.91	31.36	0.264	32.06	17.23	21.06	0.324
Purebred foreign bulls (%)	86.98	76.82	68.64	0.254	33.94	80.23	74.09	0.133

a, b Mean values with different letters in the same row are significantly different. * $p<0.05$. [1]Analysis of Variance of the groups Conventional, Organic 1 and Organic 2. [2]Analysis of Variance of the groups Conventional vs. All Organic.

Table 3. Farm breed structure. Mean percentage values, standard deviation and level of significance.

The percentage of purebred autochthonous cows reached 41.83%, with this percentage being higher in All Organic and Organic 2 than in the conventional group. Also, All Organic and Organic 2 showed a higher presence of these purebred cows; either autochthonous or foreign ones. The main reason for this is that Organic 2 farmers were market oriented (they had contracts with supermarkets) so that they knew that more productive breeds that allow them to produce carcass of better conformation, mainly Limousine.

Parameters	Conv. (*n* = 30)	Org. 1 (*n* = 22)	Org. 2 (*n* = 11)	Sig. 1[5]	Sample (*n* = 63)	SD	All Organic (*n* = 33)	Sig. 2[6]
Land fixed capital[1]	5,630.07	5,194.52	5,695.62	0.788	5,489.42	310.43	5,361.56	0.669
Buildings fixed capital[2]	660.75	606.48	546.10	0.935	621.78	114.32	586.35	0.748
Machinery fixed capital[3]	215.51	138.79	107.77	0.449	169.91	35.00	128.45	0.217
Livestock fixed capital[4]	624.81	416.66	329.10	0.217	500.49	69.69	387.47	0.089
Total fixed capital	7,131.14	6,356.45	6,678.59	0.443	6,781.59	430.59	6,463.83	0.443

Note: all these indicators were measured in terms of €/ha. [1]Value of the land at market prices. This depended on the quality of the plots (grazing resources, location and tree density, among other parameters). [2]Value of infrastructure at market prices. Years of use and level of conservation/maintenance were taken into account. [3]Value of machinery (cars, trucks, etc.) at market prices. Years of use and level of conservation/maintenance were taken into account. [4]Value of all livestock present at market prices. [5]Analysis of Variance of the groups Conventional, Organic 1 and Organic 2. [6]Analysis of Variance of the groups Conventional vs. All Organic.

Table 4. Fixed capital according to farm groups.

Parameters	Conv. (*n* = 30)	Org. 1 (*n* = 22)	Org. 2 (*n* = 11)	Sample (*n* = 63)	SD	Sig. 1[13]	All Organic (*n* = 33)	Sig. 2[14]
Feed/ha UAA[1]	109.69^{a}	17.55^{b}	96.63^{ab}	75.24	112.18	0.009**	43.91	0.019*
Feed/LU[2]	161.59^{a}	38.27^{b}	220.54^{a}	128.82	165.70	0.003**	99.03	0.136
Seeds and fertilisers[3]	7.51	3.10	1.00	4.84	12.61	0.252	2.40	0.108
Veterinary and medicines/ha UAA[4]	17.87^{a}	4.51^{b}	4.84^{b}	10.93	21.16	0.043*	4.62	0.012*
Veterinary and medicines /LU[5]	20.32^{a}	7.45^{b}	11.64^{ab}	14.31	15.14	0.006**	8.84	0.002**
Maintenance of fixed capital[6]	15.74	18.95	22.60	18.06	22.68	0.681.	20.17	0.444
Energy[7]	24.24	22.44	18.27	22.57	22.84	0.765	21.05	0.584
Other expenditure[8]	24.32	20.87	21.88	22.69	33.67	0.934	21.20	0.717
Intermediate consumption[9]	199.38	87.42	165.22	154.32	177.86	0.077	113.36	0.054
Remuneration of employees[10]	60.29	42.48	61.24	54.24	100.69	0.799	48.73	0.653
Fixed capital consumption[11]	54.59	44.20	38.08	48.08	66.62	0.744	42.16	0.464
Land rented[12]	30.56	30.96	23.25	29.42	38.47	0.846	28.39	0.825

a, b Mean values with different letters in the same row are significantly different. * $p<0.05$, ** $p<0.01$. [1]Expenditures (purchases) made in external feedstuffs/hectare of UAA (€/ha). [2]Expenditures (purchases) made in external feedstuffs/LU (€/LU). [3]Expenditure in seeds and fertilisers/hectare of UAA (€/ha). [4]Expenditures made in veterinary and medicines/ha UAA (€/ha). [5]Expenditures made in veterinary and medicines /LU (€/LU). [6]Expenditures made in maintenance of fixed capital/ha of UAA (€/ha). [7]Expenditures made in energy/ha UAA (petrol and electricity) (€/ha). [8]Expenditures/ha UAA made in farmers' associations, lawyers, etc. (€/ha). [9]Sum of the following costs (€/ha of UAA: External feedstuffs + Veterinary services and medicines + Energy + Maintenance of machinery and infrastructure + Other goods and Services (lawyers, farmers' associations, etc.). [10]Expenditures made in salaries/ha of UAA (€/ha). [11]Amortization of machinery and infrastructure = Sum of (((1/20 years amortization) × Value of infrastructures) + ((1/10 years) × Value of machinery)). [12]Cost of the land rented (€/ha). [13]Analysis of Variance of the groups Conventional, Organic 1 and Organic 2. [14]Analysis of Variance of the groups Conventional vs. All Organic.

Table 5. Intermediate consumption and other costs.

http://dx.doi.org/10.5772/65094

3.3. Economic parameters

3.3.1. *Analysis of fixed capital*

This analysis allowed identification of similarities between organic and conventional systems (**Table 4**), with regard to infrastructure, land and animals. It is worth highlighting the high average value of land fixed capital 5489.42 €/ha that accounted for the 81% of total fixed capital (**Table 5**).

Parameters (€/ha)	Conv. (n = 30)	Org. 1 (n = 22)	Org. 2 (n = 11)	Sample (n = 63)	SD	Sig.[6]	All Organic (n = 33)	Sig.[7]
Livestock sales	291.23	151.90	215.92	229.43	223.45	0.081	173.24	0.035*
Other sales	1.49	11.34	9.75	6.37	23.05	0.276	10.81	0.110
Gross output[1]	635.24	464.09	578.52	565.57	406.81	0.328	502.24	0.197
Subsidies for livestock	158.54	153.75	185.72	161.61	118.93	0.759	164.41	0.847
Total subsidies	165.70	159.91	193.85	168.59	123.39	0.752	171.22	0.861
Total income	458.41	323.15	419.52	404.39	316.03	0.313	355.27	0.198
Total subsidies/total income (%)	0.39	0.45	0.47	0.42	0.18	0.353	0.45	0.160
Net value added[2]	388.43	338.63	383.35	370.15	249.94	0.769	353.53	0.584
Net operating surplus[3]	320.98	289.98	313.99	308.94	223.74	0.886	297.98	0.687
Net entrepreneurial income[4]	290.43	259.02	290.74	279.52	214.31	0.861	269.60	0.703
Profitability rate (%)[5]	4.39	4.18	4.35	4.31	2.69	0.961	4.24	0.819

Note: Those parameters whose unit is not showed in the table are measured per ha of UAA ((€/ha). [1]Value of all the products of agricultural activities. All agricultural output was recorded except that which was solely produced by units for their own consumption. [2]It measures the value created by all agricultural output after the consumption of fixed capital. That output is valued at basic prices and intermediate consumption is valued at purchase prices. It was calculated as follows: (Gross output – Intermediate consumption – Amortisation) + (Those subsidies not related to livestock farming). [3]It measures the yield from land, capital and unpaid labour. It is the balance of the generation of income account which indicates the distribution of income between the factors of production and the general government sector. [4]Obtained by adding the interest received and then deducting rent (i.e., farm and land rents) and interest payments, measuring compensation of unpaid labour, remuneration from land belonging to units and the yield arising from the use of capital. [5]Ratio between net surplus and average capital assets, estimated from the value of total fixed capital and the value of capital. [6]Analysis of Variance of the groups Conventional, Organic 1 and Organic 2. [7]Analysis of Variance of the groups Conventional vs. All Organic.

Table 6. Economic and productive performance and subsidies.

Table 6 shows the economic and productive performance of the farm groups, as well as aspects related to subsidies, where the Organic 1 group can be seen to have lower livestock sales per hectare of UAA and lower gross production.

Conventional farms proved to sell more calves per hectare and year, which is due to their shorter productive cycle and the low productivity of Organic 1. No differences were found for the remaining indicators, but some interesting results were found and therefore comments are necessary. Organic farms (especially Organic 2) revealed higher numerical values for other sales, which reflect a higher level of business diversification, something that is key in the farms' flexibility and adaptability to the changing market environment. Moreover, organic farms (especially Organic 2) tended to be more dependent on subsidies.

4. Discussion

4.1. Structure

4.1.1. Farm and herd characteristics

All Organic farms were much smaller than the average farm size found by Perea et al. [10] in organic cattle farms located in seven regions of Spain (261.76 vs. 425 ha UAA). With regard to herd size, All Organic farms were also quite similar to Conventional farms and again smaller than the farms studied by [10], with 100.44 vs. 154 livestock units (LU).

The scarce association between land and animals continues to be an unsolved concern [6, 10]. Similarly, the integration of different livestock species is beneficial. In the farms analyzed, the proportion of cattle has been really high – 96%, in line with the findings of Perea et al. [9, 10]. This situation responds to the trend of specialisation and intensification already described [15, 17, 18], with increasing total stocking rates in beef cattle farms from 0.40 to 0.43 LU/ha ([6]—conventional farms; and [9, 10]—organic farms) to the current 0.60 LU/ha. The higher mean values observed in this study came from conventional farms (0.70). Both Organic 1 and Organic 2 farms complied with the regional organic rules [29] setting a maximum stocking rate allowed of 0.5 LU/ha.

4.1.2. Reproductive management and performance, and production results

No major differences were found between farm's groups regarding the reproductive management among groups, since most of arms followed the typical technical reproductive management in extensive ruminants production systems located in semiarid areas, where the low fertility rates compared to other breeds and systems. This is due to the fact that heats are not detected by farmers, there is no heat synchronization, and natural service is the predominant technique used for conception. Only some organic farms showed to apply artificial insemination. Average replacement rate of the sample was close to 12%, similar to that found in *dehesa* beef cattle farms, either conventional: with values ranging from 10 to 12.4% in Extremadura [30–32] or organic: 10.65% in Andalusia [9]. However, values found in the study of Milán et al. [6] were higher: 19.2%. The number of cows per bull was 30.42, lower than the 38.4 found by Milán et al. [6] and similar to the 27 found by López de Torre et al. [31] in conventional cattle farms in the *dehesas* of Extremadura. The implementation of reproductive techniques, such as artificial insemination, was even lower than that found by Milán et al. [6]: 8.5 vs. 4.80%. This divergence in results is due to the fact that they analysed farms rearing autochthonous purebred beef cattle cows. In these cases, livestock is usually registered in the Stud Book of the breeds, and the use of artificial insemination is more widespread, with the aim of rearing offspring of more appreciated genetic potential, and thus obtaining higher incomes through both selling animals as breeding animals and public subsidies.

Despite the lack of significant differences among groups, it is necessary to discuss some topics such as the reproductive calendar due to its importance in the context of uncertain availability of pastures in pasture-based systems, such as those of the Mediterranean basin. In this sense, it is recommended to avoid continuous mating and make it coincide with spring and autumn, the seasons where the availability of local feed resources allows fulfilling an important percentage of animals' nutritional needs at more affordable prices, due to a lower dependence on external feedstuff, whose prices are high and subjected to great volatility. However, also positive externalities can be found from this organization: reduced seasonality in marketing their products, thus obtaining better prices for them at certain times. Many of the farms analysed showed a distribution of mating throughout a year. Thus, the average duration of mating was 10.46 months.

Calves weaned in organic farms had lower weights than those belonging to the Conventional group. This could be due to the following aspects: Firstly, in some of the studies discussed, farms reared only local breeds, whose growing rates are lower. However, in the farms analysed in this study, many cows were either crossed or more efficient breeds, mainly the Limousine breed. Secondly, increased livestock pressure led to intensification and guidance to higher productivity which, among other adaptations, led to the inclusion of more efficient breeds. Thirdly, the rising prices of feed led to the weaning of animals at a younger age (therefore at lower weights), in order to use less feedstuff and thus reduce production costs. Finally, the next link in the food chain prefers younger animals because of their better conversion rates in feedlots. Moreover, less time grazing is usually associated with meat tenderness and lighter colour, which is in line with butchers' preferences. Thus, Organic 2 farms were those that sold more fattened calves per cow, and the age of weaning of these was lower. The latter was due to the fact that calves in Organic 2 farms were weaned before starting the fattening period, which shortened the length of the production cycle at the farm level (period between weaning and sale).

The results relating to calves weaned and sold per cow clearly show how the production of beef cattle in Southwest Europe and in semiarid areas, such as the Mediterranean basin, is mainly focused on the sale of calves at weaning. As a result, the percentage of fattened calves sold has been reduced. This is due to both the lack of infrastructure and the traditions of finishing and slaughtering animals in the Extremadura region [33]. Currently, this fact might have increased due to high feed prices and low farm profitability.

The existence of organic farms without organic products (Organic 1 in the present study) has been reported for more authors in dairy cattle [34], in a mixture of livestock and crop farms [35] and beef cattle [10, 18]. Specifically, Perea et al. [10] reported that only 40.6% of the surveyed organic beef cattle farms marketed calves as organically certified, and to the organic market. Thus, they also noticed that in different areas of Europe (from Norway to the Mediterranean area) the marketing of organic livestock is focused on the sale to conventional feedlots, and their organic stamp does not have market implications (there is a scarce market for these weaned organic animals, and they are not sold at a higher price; see [27]).

4.1.3. Breeds

The use of autochthonous breeds is a contemporary issue and usually promoted in organic farming. However, the low productivity of the rustic local cows makes it necessary to make use of other breeds that, despite not being autochthonous, are both well adapted to the local conditions and more productive. Thus, in the case of males, the racial distribution was mainly based on Limousine and Charolais breeds. This is a growing trend that responds to the need for productivity and competitiveness that requires specialisation [36]. In the *dehesas* of Extremadura, there has also been a change from Charolais towards Limousine, probably aimed at avoiding problems related to dystocia and the ability of calves to suckle, since farmers perceive that these problems are more frequent when the Charolais breed of animals are reared.

4.2. Economic parameters

4.2.1. Analysis of fixed capital

No significant differences were found between the groups of farms studied.

4.2.2. Costs, production and incomes

It is important to note that expenditure on feedstuff was lower in Organic 1 than in Conventional group when studied per hectare, while differences were found between Organic 1 and the rest of groups when these expenditures were measured per livestock unit. The expenditure on veterinary services and veterinary drugs were also lower in Organic 1 group both per area of land and per livestock unit. However, these differences only were found between Organic 1 and the Conventional group. All Organic group showed to also rely less on these external resources (feedstuff, veterinary services and drugs). However, the expenditure on feed per livestock unit was not statistically different between All Organic and the Conventional group. In general terms, these higher reliance on external resources, and in particular feed and veterinary services and drugs, is consistent with the organic production method, since the use of inputs such as feed must come from the farm itself (or the immediate surroundings), and veterinary drugs are limited to two treatments per adult cow per year, according to [28] and subsequent amendments.

When comparing Conventional and All Organic farms, one can observe very low feed costs in Organic 1 and very high feed costs in Organic 2 farms. This is due to the fact that Organic 2 farms fattened all their calves, and Organic 1, none of theirs. This increases the organic feedstuff purchased, whose price is high: around 30% above the conventional one.

The cost related to veterinary services and medicines shows that in extensive livestock systems of semiarid areas and conditions it is possible to reduce reliance on drugs with no major problems. In fact, conventional low-input farms in this area do not rely significantly on these products due to low stocking rates and dry climate. Also, as the prevalence of infectious diseases is low, it must be mentioned that the health management of organic beef cattle farms in this area is very similar to that carried out in Conventional farms, and it is not based on alternative medicine. In fact organic beef cattle farms also used some veterinary drugs as a preventive measure [17]. Organic 2 farms had higher veterinary costs than Organic 1 farms due to the fact that the transition to the fattening period usually provokes some respiratory and/or intestinal disorders.

Regarding incomes, it is necessary to increase the market orientation of Organic 1 farms, as they are not providing organic goods to the market, which influences their low economic results. Conversely, the longer productive cycle in Organic 2 farms did not allow them to clearly stand out in terms of income. Finally, the dependence on agricultural subsidies must be addressed as it is a key point for both the organic sector and the extensive beef cattle farms of Mediterranean Europe. The high dependence of this aspect makes it unstable and fragile. In the case of the organic sector, the contribution of the agro-environment subsidies makes them numerically more dependent, which is in contrast with other studies, regardless whether or not they were receiving agricultural subsidies [35, 37, 38].

The lower livestock sales per hectare of UAA and lower gross production in the Organic 1 group can be due to the fact that farms belonging to this group only sold calves at weaning age, and their prices were lower than those of fattened calves. Despite the price of organic fattened calves (marketed by the Organic farms 2 group) being greater, income from the sale of livestock per hectare of UAA was higher in Conventional farms. This was probably due to an extension of the productive cycles in Organic 2 compared to Conventional farms which, in turn, led to a reduction in the number of calves sold per cow per year. On the other hand, organic farms (especially the Organic 1 group) had higher incomes in relation to other sales (those not related to livestock). This could be a consequence of the greater degree of diversification in organic farming over conventional.

4.2.3. Other aspects worthy of discussion: workforce, agro-environmental management and marketing strategies

In addition, other aspects came up from the interviews with farmers during the farm visits which point to additional interesting aspects and open up perspectives which would be interesting to research. In this sense, Escribano et al. [17, 18] carried out a comparative sustainability assessment which showed that organic farms did not carry out so many agro-ecological practices as would be desirable to increase farm environmental protection, nutrient cycling and self-sufficiency. Moreover, these authors found that in terms of workforce both production systems are also very similar. Additionally, short marketing channels, which are commonly associated with organic production, were noted to also be very similar in various studies [17, 18, 27, 39, 40]. Profound discussion and review about these aspects can be found in other studies [40].

5. Conclusions and future perspectives

The present study integrated structure, technical, productive and economic parameters that allowed for a deep understanding of the organic beef cattle farms of Southern Europe, as well as their similarities and differences with conventional ones. Organic farms have proved to be very similar to Conventional farms (but pasture-based or low-input). Accordingly, the differences were based on the structure of the farms, more than the condition of their being organic.

According to the results discussed, it is worth mentioning that there was little orientation towards a different concept of farming, namely, environmental sustainability and self-sufficiency. However, the organic farm has been defined as a production system based on the principles of Health, Ecology, Fairness and Care. In this sense, consumers expect organic products to be based on these principles, and citizens support this system through taxes. All these aforementioned aspects shape the necessity to increase the implementation of sustainable agricultural practices, self-sufficiency and sales of organic products. Otherwise, the current production systems will hinder their sustainability due to high global competition, the increasing cost of agricultural inputs and reduced grazing resources in the Mediterranean area due to global warming. To do so, the education level of farmers, public support and farmer cooperation are essential. Moreover, further research is needed to study different production systems and strategies in order to improve the situation of the sector and the differential externalities of the organic livestock sector above the conventional one.

Acknowledgements

Research funded by project INIA-RTA2009-00122-C03-03 of the Spanish Ministry Economy and Competitiveness. The authors would also like to acknowledge the support provided by Junta de Extremadura and FEDER funds. Special thanks to the farmers, practitioners and experts that contributed to this research. The first author also acknowledges the pre-doctoral financial support of the Fundación Fernando Valhondo Calaff

Author details

Alfredo J. Escribano[1*], Paula Gaspar[2], Francisco J. Mesías[3] and Miguel Escribano[4]

*Address all correspondence to: ajescc@gmail.com

1 C/Rafael Alberti, Cáceres, Spain

2 Department of Animal Production and Food Science, Faculty of Agriculture, University of Extremadura, Ctra. Cáceres s/n, Badajoz, Spain

3 Department of Economy, Faculty of Agriculture, University of Extremadura, Ctra. Cáceres s/n, Badajoz, Spain

4 Department of Animal Production and Food Science, Faculty of Veterinary Sciences, University of Extremadura, Avda.de la Universidad s/n. Cáceres, Spain

References

[1] FiBL, IFOAM. The world of organic agriculture. Statistics & Emerging Trends 2015. 2015; http://www.organic-world.net/yearbook-2015.html. Accessed 04 August 2015.

[2] Rodríguez L, Bartolomé DJ, Tabernero de Paz MJ, Posado R, García JJ, Bodas R, Garrido C, Vázquez JM, Mazariegos V, Vicente M, Olmedo S. Características técnico-económicas de las explotaciones de ovino lechero con reproducción asistida de Castilla y León: Sistemas y tipos de explotación. ITEA. 2013;109:201-214.

[3] Olaizola AM, Manrique E, Maza MT. Tipos de sistemas de producción y rendimientos económicos en las explotaciones de vacuno de montaña. ITEA Producción Animal. 1995;91:47-58.

[4] Serrano EP, Lavín P, Giradles FJ, Bernués A, Ruiz A. Classification variables of cattle farms in the mountains of León, Spain. Spanish Journal of Agricultural Research. 2004;2:504–511.

[5] Serrano E, Lavín P, Giradles FJ, Bernués A, Ruiz A. The identification of homogeneous groups of cattle farms in the mountains of León, Spain. Spanish Journal of Agricultural Research. 2004;2:512-523.

[6] Milán MJ, Arnalte E, Caja G. Economic profitability and typology of Ripollesa breed sheep farms in Spain. Small Ruminant Research. 2003;49:97-105.

[7] Nieto MI, Olaizola AM, Manrique E. Los sistemas de vacuno de carne de las sierras de La Rioja: I Caracterización de las explotaciones. ITEA Producción Animal. 2005;26:183-185.

[8] Ruiz R, Igarzabal A, Mandaluniz N, Amenazar M, Oregui LM. Caracterización de los principales sistemas de producción de ganado vacuno en el País Vasco. ITEA. 2005;26:195-197.

[9] Perea J, García A, Acero R, Valerio D, Rodríguez V. Caracterización productiva del vacuno ecológico en Andalucía. Archivos de Zootecnia. 2007;56:517-521.

[10] Perea J, Blanco–Penedo I, Barba C, Angón E, García A. Organic beef farming in Spain: Typology according to livestock management and economic variables. (Producción ecológica bovina de carne en España: tipología de acuerdo a variables técnicas y económicas). Revista Científica, FCV-LUZ. 2014;XXIV:347-354.

[11] Granda M, Moreno V, Prieto M, (Eds.) Pastos naturales en la dehesa extremeña. SIA-Junta de Extremadura. 1991.

[12] Hernández CG, (Ed.) La dehesa, aprovechamiento sostenible de los recursos naturales. Editorial Agrícola Española. 1998.

[13] Espejo M, Espejo AM. Los sistemas tradicionales de explotación y la aplicación de innovaciones tecnológicas de la dehesa. In: Espejo M, Martín M, Matos C, Mesías FJ,

(Eds.) Gestión ambiental y económica del ecosistema dehesa en la Península Ibérica. Junta de Extremadura, Consejería de Infraestructuras y D.T. 2006.

[14] Gaspar P, Mesías FJ, Escribano M, Rodríguez De Ledesma A, Pulido F. Economic and management characterization of dehesa farms: Implications for their sustainability. Agroforestry Systems. 2007;71:151-162.

[15] Gaspar P, Mesías FJ, Escribano M, Pulido F. Sustainability in Spanish extensive farms (Dehesas): An economic and management indicator-based evaluation. Rangeland Ecology and Management. 2009;62:153-162.

[16] Franco JA, Gaspar P, Mesías FJ. Economic analysis of scenarios for the sustainability of extensive livestock farming in Spain under the CAP. Ecological Economics. 2012;74:120-129.

[17] Escribano AJ, Gaspar P, Mesías FJ, Escribano M, Pulido F. Comparative sustainability assessment of extensive beef cattle farms in a high nature value agroforestry system. In: Squires VR, (Ed.) Rangeland Ecology, Management and Conservation Benefits. Nova Science Publishers. 2015.

[18] Escribano AJ, Gaspar P, Mesías FJ, Pulido AF, Escribano M. A sustainability assessment of organic and conventional beef cattle farms in agroforestry systems: The case of the dehesa rangelands. ITEA Información Técnica Económica Agraria. 2014;110:343-367.

[19] Regulation (EC) No 138/2004 of the European Parliament and of the Council of 5 December 2003 on the economic accounts for agriculture in the Community.

[20] Gaspar P, Escribano M, Mesías FJ, Ledesma AR, Pulido F. Sheep farms in the Spanish rangelands (dehesas): Typologies according to livestock management and economic indicators. Small Ruminant Research. 2008;74:52-63.

[21] Lebacq T, Baret PV, Stilmant D. Sustainability indicators for livestock farming. A review. Agronomy for Sustainable Development. 2013;33:311-327.

[22] Nieto MI, Olaizola AM, Manrique E. Los sistemas de vacuno de carne de las sierras de La Rioja: I Caracterización de las explotaciones. ITEA. 2005;26:183-185.

[23] García A, Perea J, Acero R, Angón E, Toro P, Rodríguez V, Gómez Castro AG. Structural characterization of extensive farms in andalusian dehesas. Archivos de Zootecnia. 2010;59:577-588.

[24] Gaspar P, Escribano AJ, Mesías FJ, Escribano M, Pulido AF. Goat systems of Villuercas-Ibores area in SW Spain: Problems and perspectives of traditional farming systems. Small Ruminant Research. 2011;97:1-11.

[25] Toro-Mújica P, García A, Gómez-Castro AG, Acero R, Perea J, Rodríguez-Estévez V, Aguilar C, Vera R. Organic dairy sheep farms in south-central Spain: Typologies

according to livestock management and economic variables. Small Ruminant Research. 2012;100:89-95.

[26] Riveiro JA, Mantecón AR, Álvarez CJ, Lavín P. A typological characterization of dairy Assaf breed sheep farms at NW of Spain based on structural factor. Agricultural Systems. 2013;120:27-37.

[27] Escribano AJ. Beef cattle farms' conversion to the organic system. Recommendations for success in the face of future changes in a global context. Sustainability. 2016;8:572.

[28] Regulation (EC) No 834/2007 of 28 June 2007 on organic production and labelling of organic products and repealing Regulation (EEC) No 2092/91.

[29] DECRETO 9/2016, de 26 de enero, por el que se regula el régimen de ayudas agroambientales y climáticas y agricultura ecológica para prácticas agrícolas compatibles con la protección y mejora del medio ambiente en la Comunidad Autónoma de Extremadura. DOE Numero 18. Jueves, 28 de enero de 2016.

[30] López de Torre G.. Metodología y esquema de selección de razas maternales de vacuno de carne: Aplicación a la raza Retinta. Especial Vacuno de Carne. Exclusivas ONE. 1986:75-93.

[31] López de Torre G, García Barreto L, López Márquez JA. Encuesta sobre la explotación y mejora del ganado vacuno en la Dehesa extremeña. In: Campos P, Martín M, (Eds.) Conservación y Desarrollo de las Dehesas Portuguesas y Española. Ministerio de Agricultura, Pesca y Alimentación. 1987.

[32] Martín M, Escribano M, Mesías FJ, Rodríguez de Ledesma A, Pulido F. Sistemas extensivos de producción animal. Archivos de Zootecnia. 2001;50:465-489.

[33] Mateos A. 2008. Caracterización del cebo de bovino en Extremadura. Doctoral thesis. Departamento de Producción Animal, Universidad de Extremadura.

[34] Flaten O, Lien G, Ebbesvik M, Koesling M, Valle PS. Do the new organic producers differ from the 'old guard'? Empirical results from Norwegian dairy farming. Renewable Agriculture & Food Systems. 2006;21:174-182.

[35] Argyropoulos C, Tsiafouli MA, Sgardelis SP, Pantis JD. Organic farming without organic products. Land Use Policy. 2013;32:324-328.

[36] Ligios S, Revilla R, Nardona A, Casu S. Cattle husbandry systems in Mediterranean mountains: Situation and prospects. In: Georgoudis A, Rosati A, Mosconi C, (Eds.) Animal Production and Natural Resources Utilisation in the Mediterranean Mountain Areas EAAP. 2005.

[37] Hrabalová A, Zander K. Organic beef farming in the Czech Republic: Structure, development and economic performance. Agricultural Economics UZPI. 2006;52:89-100.

[38] Gillespie J, Nehring R. Comparing economic performance of organic and conventional U.S. beef farms using matching examples. Australian Journal of Agricultural and Resource Economics. 2013;57:178-192.

[39] Lobley M, Butler A, Winter M. Local organic food for local people? Organic marketing strategies in England and Wales. Regional Studies. 2013;47:216-228.

[40] Escribano AJ. Organic livestock farming: Challenges, perspectives, and strategies to increase its contribution to the agrifood system's sustainability – A review. In: Konvalina P, (Ed.) Organic Farming – A Promising Way of Food Production, InTech. 2016.

Chapter 3

Naturalized Breeds in Brazil: Reports on the Origin and Genetic Diversity of the Pantaneiro Sheep

Bruno do Amaral Crispim,
Alexandre Campos Banari,
Joyce Azambuja de Oliveira,
Juliana dos Santos Fernandes and
Alexeia Barufatti Grisolia

Additional information is available at the end of the chapter

http://dx.doi.org/10.5772/66019

Abstract

Brazil has several breeds of sheep, including animals that developed from breeds brought by settlers soon after their discovery. Over the years, these animals were under the process of natural selection, resulting in breeds that are considered naturalized. The Pantaneiro sheep shows rusticity and ability of adaptation to tropical climate regions and tolerance or resistance to disease and parasites. Molecular tools have marked the discovery of the origin and domestication processes of a wide variety of species, using both nuclear and mitochondrial molecular markers. These tools have aided in the understanding of evolutionary relationships, taxonomies, and demographics of various species and provided support to identify the most important areas for conservation programs, in addition to assisting in the analysis of genetic diversity in domestic, wildlife and endangered species. Researches using these tools show the importance of exploiting the potential of the genetic diversity found in locally adapted livestock. So far, a few studies were performed to observe that Pantaneiro sheep served as maternal basis for the origin of other breeds reared. Moreover, it is possible to suggest an European origin for the sheep populations studied; therefore, more studies using more markers are needed, so that it is possible to prove their origin.

Keywords: Pantaneiro sheep, genetic management, phylogeny, mtDNA, nuclear DNA

1. Introduction

Brazil has several breeds of sheep, including animals that developed from breeds brought by settlers soon after the discovery. Over the years, these animals were under the process of natural selection of local environmental and climatic conditions, resulting in breeds that are considered naturalized, locally adapted or native [1].

Thus, the characterization of the diversity of naturalized breeds, the genetic relationship between them, as well as knowledge of their origins in other breeds are the first step to obtain subsidies for breeding programs, management, and conservation of naturalized Brazilian sheep [2].

Molecular tools and recent technologies have marked the discovery of the source and domestication processes of species, both using molecular markers such as mitochondrial DNA (mtDNA). These tools have aided the understanding of evolutionary relationships, taxonomies, demographics of various species and provided support to identify the most important areas for conservation programs, and assist in the analysis of genetic diversity in domestic animals, wildlife, and endangered species [3, 4].

The mtDNA shows the haplotype diversity within species; therefore, it becomes a useful tool to establish phylogenetic relationships between the species [5]. On the other hand, the haploid inheritance of markers of the Y chromosome makes them to be extremely sensitive for detecting genetic history, the process of domestication, the relationship between population and the male abundance in gene flow [6].

The Pantaneiro sheep showed greater allelic richness when compared with other populations in a study with other six breeds in the state of Mato Grosso do Sul, Brazil [7]. Considering that the introduction of sheep in South America has occurred concomitant with the process of colonization and the effective population size is in process of formation, high levels of diversity observed in the different studied populations can be related to the fact of selective pressure coupled with miscegenation have implicated in the introgression of genes in populations.

Analyzing the haplotypes found in mtDNA, it was observed that the Pantaneiro sheep was distributed in seven haplotypes and grouped with all other locally adapted breeds that were analyzed [8].

A study performed with Creole breed demonstrated that these animals have a different haplotype compared to the animals of Pantaneiro breed, as there was no grouping in the same network [9]. This suggests a difference between these groups, so more research would be needed to see if it is enough that the Pantaneiro sheep can be recognized as a distinct breed [10]. However, it was also observed that several haplotypes for the Creole breed were close to haplotype formed by Pantaneiro breed which could indicate that, although these animals do not share the same haplotype, they belong to the same haplogroup. The regions where these animals currently found previously belonged to Paraguay, so it is possible that Pantaneiro breed has been influenced by Paraguayan herds. Analysis of their mitochondrial genome can redeem this history showing their genetic diversity from the Creole breed. Thus, the existence of significant differences in the ND5 gene of mtDNA between Pantaneiro and Creole breeds

http://dx.doi.org/10.5772/66019

could indicate that differentiation has occurred between the two breeds, but further research using other markers would be needed to prove the differentiation between Pantaneiro and Creole.

Research shows the importance of exploiting the potential of the genetic diversity found in locally adapted livestock. Innate genes with adaptive values linked to tropical climatic regions compared to breeds created/improved in temperate environments could be highly useful in view of the climate changes predicted for the coming years. The exotic breeds, although considered high performance, can reduced their productivity by not easily conform to the conditions of creation and management taxes in Brazil, combined yet to climatic weather (tropical). The introgression of genes between these populations can produce animals whose general average production and rusticity exceed the means of the parents.

A few studies demonstrated that Pantaneiro breed served as maternal basis for the formation of other breeds reared in the region [8, 11]. Moreover, it is possible to suggest an European origin for the sheep populations studied; therefore, more studies using more markers are needed so that it is possible to prove their origin.

It is assumed therefore that the establishment of conservation programs using molecular tools is crucial to provide information regarding the genetic diversity patterns of locally adapted groups, and allow the same to be used for the production system, adding adaptation features, and rusticity [12].

2. Brazilian naturalized sheep

Domestic sheep are belonging to the Bovidae family, *Ovis* genus and *Ovis aries* species. The sheep were the first animals to be domesticated, and it is believed that its origin is descended from the Muflon (*Ovis musimon*) and Urial (*Ovis orientalis*) breeds, since the Urial breed may have contributed to the formation of all European sheep breeds [13].

Brazilian sheep breeds, as well as several species of domestic animals in the country were introduced by Portuguese colonization. Over the years, they have been subjected to natural selection because of their adaptation and survival to the local environmental and climatic conditions, resulting in breeds that are now considered locally adapted or local, creole and native or naturalized [1, 2].

Brazilian naturalized sheep are small and specialized in intensive meat or milk production system, so far they have not undergone to selection process and genetic breeding [2]. These breeds were known for their rusticity and ability of adaptation to tropical and subtropical climate regions, allowing them to hold important attributes of genetic resources. These animals still have tolerance or resistance to disease and parasites, as well as adaptation to availability of food resources and water [14].

However, the search for more productive breeds from the ending of the nineteenth century and the beginning of the twentieth century led to import exotic breeds from temperate climatic region, which did not have the adaptive traits of local breeds. This way, it occurred indiscriminate crossings, which was determinant to result in rapid replacement and erosion of naturalized animals, endangering their existence [15].

In order to prevent the extinction of naturalized breeds and their genetic heritage, in 1983, the National Research Center for Genetic Resources and Biotechnology (Cenargen) of the Brazilian Agricultural Research Corporation (Embrapa) decided to include animal genetic resource conservation research in its Genetic Resources Conservation program (Brazilian Animal Genetic Resources Conservation program) [16]. Since then, in addition to Cenargen, several Embrapa centers, universities, private companies, and farmers have maintained the conservation of animals, through natural selection and semen, embryos, and oocytes storage in germplasm banks [17].

3. Pantaneiro sheep

The Brazilian Pantanal, considered the largest wetland sedimentary plain of the world, is located in the states of Mato Grosso and Mato Grosso do Sul and part of the northern Paraguay and eastern Bolivia. This biome, due to its intrinsic characteristics, made it difficult the exchange for other regions of the country at the time of colonization. So, the first animal's populations developed for breeds and the sheep of Pantanal region is an example of adaptability maintained throughout the years [18].

Figure 1. Pantaneiro sheep.

With the expansion of sheep breeding in the state of Mato Grosso do Sul, the genetic group of locally adaptaded sheep, known as Pantaneiro Sheep (**Figure 1**), can be advantageous to increase the production chain, due to its adaptation to the environmental conditions of the region and by using its rustic genetic material in conservation and breeding programs [7].

In 2005, researchers at the University Anhanguera (UNIDERP), Federal University of Grande Dourados (UFGD), the Brazilian Agricultural Research Corporation (EMBRAPA), and the Federal University of Mato Grosso do Sul (UFMS) initiated studies with the Pantaneiro genetic sheep group in order to identify, preserve, record, and ensure the development of animals and their desirable traits obtained by natural selection. The groups were consisted of animals obtained from farms of Pantanal (Midwest of Brazil), which had similar phenotypic traits but distinct genotypic patterns presented by exotic breeds created in Brazil [7, 18, 19].

4. Features of Pantaneiro sheep

The Pantaneiro sheep is a separate genetic group of sheep breeds, and it presents allelic combination approaching to wool breeds from South Brazil and woolless breeds from Northeast Brazil [20]. As the Pantaneiro sheep are living in the Pantanal region for many years without suffering any kind of artificial selection or genetic breeding, it was concluded that these animals are locally adapted to the region [19]. Evidence of this adaptability can be determined through the wool distribution in the body of the animals, since they show little or no wool in the legs, belly, and neck, once these body parts are in the most contact with water, when there is need for locomotion in local filled with water and dense vegetation [18].

In a biometric analysis of 20 Pantaneiro sheep, lower morphometric data were observed, but similarity exists between native animals of the Pantanal and other sheep breeds created in Brazil. However, the heart girth and rump width measurements were higher in Pantaneiro sheep than the data in the literature for crossing between Ile de France and Ideal breeds. Therefore, the Pantaneiro animals show good potential for genetic breeding sheep breeding [21].

For assessing the morphometric measurements, a study with four lambs obtained the following averages for the characteristics studied: body weight (46.3 kg), body (69.6 cm), and croup length (19, 1 cm), chest circumference (82.7 cm), chest (18.1 cm), and croup width (11.7 cm). The mean values for these characteristics were smaller when compared to the values observed in wool Creole sheep, but they were higher when confronted with three woolless breeds (Santa Inês, Morada Nova, and Brazilian Somalis) [22].

An analysis of morphometric characterization, using 338 naturalized sheep from Mato Grosso do Sul and performed 20 morphometric measurements, concluded that the females of this genetic group have relatively large heads (32.4 cm), and this feature along with the length of the skull (24.2 cm) are larger in females than in males [23].

The Pantaneiro sheep have good productive potential in relation to meat and carcass quality traits, and also for the production of wool, widely used in handicrafts [19]. The lean physique of these animals owing to their nonrequirement for the high-calorie consumption results in low subcutaneous fat accumulation, which is characterized by their rusticity [24].

These animals exhibit average birth weight between 2.5 and 3.5 kg and about 200 to 350 g in average daily gain weight. Still, it was found that the weight data support the production of lambs, providing slaughtering when the animals are between 4 and 8 months, weighing between 30 and 40 kg, generating high carcass yield, up to 50%, and in addition males and females have similar productive performance [19].

These sheep also features highlighted in relation to the reproductive performance when compared to other breeds of sheep created in Brazil. With reproductive activity during the entire year, females do not show reproductive seasonality, and they can produced more than 1.5 lamb per year, unlike the females of other breeds [19, 25, 26].

The male sheep also have reproductive seasonality, but only when there is little variation of photoperiod, as in the tropics. Moreover, the variation of breed and environmental factors are also crucial to the reproductive performance of animals [19].

As Pantaneiro sheep are not considered a breed, studies of molecular, genetic, and phenotypic characterization are needed in order to prove the differences between these animals and the other breeds. More studies are required to demonstrate the body, production and reproduction characteristics of these animals [27].

5. Genetic diversity

Through indiscriminate crossbreeding with exotic breeds, mainly imported from Africa and Europe, it was put at risk the existence and preservation of naturalized breeds that are now important genetic heritage [28]. These animals have characteristics which may be considered useful both from the productive point of view and adaptive such as tolerance or resistance to diseases and parasites and adaptive wide variations related to the availability and quality of food and water. Therefore, the animals best adapted and/or more resistant survived and reproduced to the present day. Thus, the naturalized breeds are a result of the process of natural selection over the years.

Research aimed at the conservation and breeding of naturalized breeds that are important for the selection of animals for the purpose of controlling intersections, avoiding inbreeding, and indiscriminate crossbreeding that thus purebreds are preserved. Therefore, it is necessary to seek a production system that makes evident its potential, so that they are recognized by the creators and that they perceive the possibility of the use of local breeds for higher returns [29].

Studies related to knowledge of adaptive characteristics of different breeds to the environment can sustain production systems in livestock based on adapted breeds, reducing the impact on the environment and receiving better products for consumption.

With recent technological developments, new molecular tools were developed in order to understand the origin and domestication process of domestic species. These tools helped the discovery of evolutionary relationships, taxonomy, and demographics of a wide variety of species, providing important support both in identifying priority areas for conservation programs and in the understanding of genetic diversity in domestic and wild species threatened with extinction [3].

6. Microsatellite markers

Microsatellites are the most widely used marker for the study of genetic diversity and population structure of domestic animals [30]. The abundance of this marker along the genome, its high degree of polymorphism and codominance are the main features that make it an important tool for genomic analysis.

Through the evaluation of eight microsatellite loci in five breeds of unrelated sheep (Romney, Border Leicester, Suffolk, Awassi, Australian, and New Zealand Merino) was found highly significant differences in allele frequencies between individuals, indicating that the genotyping using microsatellite can be a useful tool for examining the evolutionary relationships between the breeds [31]. Studies using microsatellite markers to characterize genotypic and assessments of genetic diversity of sheep described in Spanish breeds, determined the genetic relationship between Swiss breeds. In these studies, microsatellites were efficient to evaluate genetic diversity and demonstration of the genetic diversity between the animals involved [32, 33].

Several studies have shown the use of microsatellite markers in genetic diversity studies of native animals of Brazil. In a study using 27 microsatellite markers to analyze the genetic variability of native breeds of goats in Brazil, the result showed that all microsatellites were polymorphic and showed a high capacity for genetic characterization of these breeds [34]. In another study, microsatellite markers for 18 loci in studies of genetic diversity of sheep naturalizes and exotic in Brazil were used (Santa Inês, Bergamácia Wide Tail, Morada Nova and Somali) and the results showed the efficiency of these markers in the characterization of these breeds because all breeds differed significantly, although they presented low genetic variability [2].

A study with 717 animals was determined the variability of 20 microsatellites in 14 Portuguese sheep breeds. Analysis of these results allowed us to assess the degree of structure of the Portuguese population of sheep and estimate parameters of genetic diversity in each of the breeds [35].

Thus, microsatellites have proven marker of excellence for characterization of new naturalized breeds [36], as well as to genetic variability of population studies [37, 38]. A recent study identified an approach to facilitate the merger of microsatellite data for cross-country comparison of genetic resources when samples are evaluated in different laboratories. This approach can facilitate the merger and analysis of microsatellite data for cross-country comparison and extend the utility of previously collected molecular markers. In addition, this analysis can be used in new and existing conservation programs [39].

Recent research analyzes genetic diversity and population structure among varieties of sheep. Therefore, nuclear microsatellite markers and regions of mitochondrial DNA are used [40].

7. Phylogeny and population structure

Despite the sheep breeds are considered adapted to Brazil and these animals were brought to the country by settlers soon after the discovery, few studies have been conducted in order to discover the origin of these animals.

Knowledge of the population structure combined with information about genetic changes can influence future management actions and can be used to develop strategies for using a breed in a particular ecosystem as a model for genetic improvement programs [41]. Conservation programs using molecular tools are crucial for the providing information about the genetic diversity of locally adapted groups, thus allowing them to be included in production systems for integration of adaptation and rusticity features [12].

Recent molecular tools and technologies have marked the discovery of the origin and domestication processes of a wide variety of species, using either nuclear or mitochondrial molecular markers. These tools have aided in the understanding of evolutionary relationships, taxonomy, and demography of several species that will provide support to identify the most important areas for conservation programs, in addition to the analysis of genetic diversity in domestic, wildlife, and endangered species [3, 4].

Two studies [42, 43] demonstrated the existence of at least two major haplogroups in *O. aries* from the control region (D-loop) of mitochondrial DNA (mtDNA) sequencing: one of European origin and another, probably of Asian origin. These results can also be interpreted as two independent domestication events that have occurred for domestic species [44]. Furthermore, it was developed a test based on polymerase chain reaction—restriction fragment length polymorphism (PCR-RFLP) of mitochondrial cytochrome C oxidase I gene (MT-COI 6) with the restriction enzyme *Hinf*I (extracted from bacteria *Haemophilus influenza* Rf) in order to more easily identify these two haplogroups HA (Asian origin) and HB (European origin) [43].

The study of the mtDNA region, which can be called DNA barcoding, uses partial DNA sequences of the MT-COI 6 gene to identify and designate both new species as described previously, helping to unravel the diversity [45].

A study using PCR-RFLP from MT-COI 6 gene using HinfI restriction enzyme was performed to molecularly characterize, over the existing haplogroups, some sheep breeds used in the state of Mato Grosso do Sul [8]. The study with the MT-COI RFLP gene indicated the applicability of this molecular tool to classify most of the animals as belonging to the European haplogroup, highlighting the European origin of the state breeds.

Researchers analyze genetic diversity and population structure among varieties of White, Red, and Black Morada Nova hair sheep from flocks in the northeastern Brazilian semiarid region. In this study, the use of 15 nuclear microsatellite markers and two regions of mitochondrial DNA identified the existence of substantial differences between the Red and White varieties of this sheep and should be used as separate genetic resources and to improve conservation programs [40].

The origin of sheep from some of the breeds in the state of Mato Grosso do Sul, Brazil is important because these are part of the genetic heritage of the state and by knowing their phylogeny it is possible to improve the management of these breeds, aiming its conservation and the use of the productivity of these animals in our environment.

The Creole sheep has been reared for centuries in the Brazilian states of Rio Grande do Sul and Santa Catarina, where there are two known varieties: Fronteira and Serrana [46]. Considering the geographic distribution of sheep in Brazil and phenotypic similarities between the animals, it is thought that Pantaneiro sheep originated from the Creole sheep, and research has been carried out to determine whether the difference between the groups is sufficient for the Pantaneiro sheep to be recognized as a separate breed [10]. NADH dehydrogenase is one of the main enzymes found in respiratory complexes in mammals. The subunit five (*ND5*) was used to study sheep diversity [47, 46]. The former study determined subspecies of *Ovis ammon* in Mongolia by sequencing this region, and the results suggested the existence of two subspecies (*O. ammon ammon* and *O. ammon darwini*). The genetic differentiation was found between animals of the Creole sheep in the south of Brazil belonging to the varieties Serrana and Fronteira [44].

Genetic polymorphisms in mitochondrial DNA (mtDNA) reveal haplotype diversity within species, and are therefore a useful tool for establishing phylogenetic relationships at the species level [5] The Pantaneiro breed presents a higher genetic variability than other breeds of sheep reared in tropical altitude regions. Therefore, it is important to develop research that aims at their conservation and genetic improvement [7].

In order to assessed the variation between a population of Pantaneiro sheep in the state of Mato Grosso do Sul and Creole sheep in the south of Brazil through molecular analysis of the mtDNA ND5 region, an study revealed that Creole sheep have a different haplotype compared to Pantaneiro sheep, suggesting that differentiation has occurred between these groups; therefore, more research would be necessary that can be recognized Pantaneiro sheep as a distinct breed [9, 10]. Furthermore, several haplotypes in the Creole sheep were close to the one formed by the Pantaneiro breed animals which may indicate that, although these animals do not share the same haplotype, they belong to the same haplogroup. The geographical region where these animals are found today belonged previously to Paraguay, so it is possible that the Pantaneiro breed has been influenced by Paraguayan breeds and the analysis of its mitochondrial genome might confirm this assumption, by showing their genetic diversity from the Creole.

Thus, the significant differences identified for the mtDNA *ND5* gene between Pantaneiro and Creole sheep may indicate that differentiation has occurred in both breeds; however, further research using other markers is required to investigate this further. Additional management measures need to be carried out in this herd to reduce inbreeding and optimize genetic variation.

Other aspects besides the distribution of genetic diversity have to be taken into account when dealing with conservation strategies of species. Historical, cultural, and traditional aspects about the use of particular breeds are relevant issues. Furthermore, the selection practiced by sheep breeders can favoring alleles for which the surrogate neutral markers used in diversity surveys are not necessarily fully representative.

Author details

Bruno do Amaral Crispim, Alexandre Campos Banari*, Joyce Azambuja de Oliveira, Juliana dos Santos Fernandes and Alexeia Barufatti Grisolia

*Address all correspondence to: alexeiagrisolia@ufgd.edu.br

Faculty of Biological and Environmental Science, Federal University of Grande Dourados, Dourados, Brazil

References

[1] Mariante, A.S., Albuquerque, M. S.M., Egito, A. A., McManus, C. Advances in the Brazilian animal genetic resources conservation programme. Animal Genetic Resources Information. 1999;25:109–123. DOI: http://dx.doi.org/10.1017/S1014233900003497.

[2] Paiva, S.R., Silvério, V.C., Paiva, D.A.F., McManus, C., Egito, A.A., Mariante, A.S., et al. Origin of The main locally adapted sheep breeds of Brazil: A RFLP-PCR Molecular Analysis. Arch. Zootec. 2005a;54:395–399.

[3] Rosa, A.D.M., Paiva, S. Molecular markers and their application in population studies of livestock species. Planaltina, DF:Embrapa Cerrados, Documents; 2009; 254. 35 p.

[4] Grisolia, A.B., Moreno-Cotulio, V.R. Molecular Markers and Genetic Diversity in Neotropical Felids. In: Mahmut Caliskan, editor. Analysis of Genetic Variation in Animals. Rijeka:InTech; 2012. p. 105–120. DOI: 10.5772/34103. Available from: http://www.intechopen.com/books/analysis-of-genetic-variation-in-animals/molecular-markers-and-genetic-diversity-in-neotropical-felids.

[5] Avise, J.C., Arnold, J., Ball, R.M., Bermingham, E., Lamb, T., Neigel, J.E., Reeb, C.A., Saunders, N.C. Intraspecific phylogeography: the mitochondrial DNA bridge between population genetics and systematics. Ann. Rev. Ecol, Evol. Systemat. 1987;18:489–522. DOI: 10.1146/annurev.es.18.110187.002421.

[6] Oner, Y., Calvo, J., Elmaci, C. Y. chromosomal characterization of Turkish native sheep breeds. Livest. Sci. 2011;136(2–3): 277–280. DOI: 10.1016/j.livsci.2010.08.015.

[7] Crispim, B.A., Grisolia, A.B., Seno, L.O., Egito, A.A., Vargas Junior, F.M., Souza, M.R. Genetic diversity of locally adapted sheep from Pantanal region of Mato Grosso do Sul. Genet. Mol. Res. 2013;12(4):5458–5466. DOI: 10.4238/2013.

[8] Oliveira, J.A., Crispim, B.A., Banari, A.C., Egito, A.A., Vargas Junior, F.M., Seno, L.O., et al. Segregation of MT-COI RFLP in sheep from Mato Grosso do Sul, Brasil. Arch. Zootec. 2016;65(250):231–233.

[9] Oliveira, J.A., Egito, A.A., Crispim, B.A, Vargas Junior, F.M., Seno, L.O., Grisolia, A.B. Analysis of polymorphisms in the mitochondrial ND5 gene in Pantaneira and Creole breeds of sheep. Afr. J. Biotechnol. 2015;14(5):438–441. DOI: 10.5897/AJB2014.14284.

[10] Paiva, S.R.; Santos, S.A.; Vilarinho, K.R. Origin and genetic diversity of the Creole sheep of the Pantanal-MS region. In: Partial Research Report Concerning to the Research Activities between Embrapa Pantanal and Embrapa Genetic Resources and Biotechnology. Brasília:Embrapa; 2008.

[11] Crispim, B.A., Seno, L.O., Egito, A.A., Vargas-Junior, F. M., Grisolia, A.B. Application of microsatellite markers for breeding and genetic conservation of herds of Pantaneiro sheep. Electron. J. Biotechnol. 2014; 17(6):317–321. DOI: 10.1016/j.ejbt.2014.09.007.

[12] Barker, J.S.F. A global protocol for determining genetic distances among domestic livestock breeds. In: Proceedings of the 5th World Congress on Genetics Applied to Livestock Production. Guelph; 7–12 August 1994; Ontario. Guelph, Ontario, Canada: University of Guelph; 1994. p. 501–508.

[13] Scherf, B.D. World watch list for domestic animal diversity. 3rd ed. Roma: FAO; 2000. 726 p.

[14] Crispim, B.A., Matos, M.C., Seno, L.O., Grisolia, A.B. Molecular markers for genetic diversity and phylogeny research of Brazilian sheep breeds. Afr. J. Biotechnol. 2012;25(90):15617–15625. DOI: 10.5897/AJB12.2275.

[15] Egito, A.A., Mariante, A.S., Albuquerque, M.S.M. Programa brasileiro de conservação de recursos genéticos animais. Arch. Zootec. 2002;51:39–52.

[16] Mariante, A.S., Egito, A.A., Albuquerque, M.S.M., Paiva, S.R., Ramos, A.F. Managing genetic diversity and society needs. R. Bras. Zootec. 2008;37:127–136. DOI: 10.1590/S1516-35982008001300016.

[17] Mariante, A.S., Egito, A.A. Animal genetic resources in Brazil: result of five centuries of natural selection. Theriogenology. 2002;57(1): 223–235. DOI: 10.1016/S0093-691X(01)00668-9.

[18] Costa, J.A.A., Egito, A.A., Barbosa-Ferreira, M., Reis, F.A., Vargas Junior, F.M., Santos, S.A., et al. Pantaneiro Sheep, a genetic group naturalized from the state of Mato Grosso do Sul, Brazil. In: Latin American congress of specialists in South American small Ruminants and camelids; 16–18 May. Campo Grande, MS:Embrapa; 2013. p. 25–43.

[19] Vargas Junior, F.M., Longo, M.L., Seno, L.O., Pinto, G.S., Barbosa-Ferreira, M., Oliveira, D.P. Productive potential of a genetic group of Sulmatogrossenses native sheeps. PUBVET. 2011a;5(30):1197.

[20] Gomes, W.S.G., Araújo, A.R., Caetano, A.R.; Martins, C.F., Vargas Junior, F.M., McManus, C., et al. Origin and genetic diversity of the Pantanal Creole Sheep, Brazil. In: Symposium of Genetic Resources for Latin America and the Caribean. Chapingo, México:Universidad Autónoma Chapingo; 2007.

[21] Vargas Junior, F.M., Martins, C.F., Souza, C.C., Pinto, G. S., Pereira, H.F., Camilo, F.R., et al. Biometric Evaluation of Pantaneiros Lamb. Agrarian. 2011b;4(11):60–65.

[22] Carmona, R. Sheeps' morphometry of Pantaneiro genetic group from CTO [thesis]. Campo Grande, MS: Anhanguera-UNIDERP University; 2011. 43 p. Available from: http://docplayer.com.br/16213295-Universidade-anhanguera-uniderp-programa-de-mestrado-profissional-em-producao-e-gestao-agroindustrial-reinaldo-carmona.html.

[23] Oliveira, D.P., Oliveira, A.L., Martins, E.N., Vargas Junior, F.M., Barbosa-Ferreira, M., Seno, L.O., et al. Morphostructural characterization of female and young male of naturalized Sul-mato-grossenses "Pantaneiros" sheep. Semina: Ciências Agrárias. 2014;35(2):973–986. DOI: 10.5433/1679-0359.2014v35n2p973.

[24] Ferreira, M. Internet]. 11/01/2011. Available from: http://ruralcentro.uol.com.br/analises/2214/resumo-historico-do-ovino-pantaneiro [Accessed: 08/13/2016].

[25] Martins, C.F., Vargas Junior, F.M., Pinto, G.S., Nogueira, M.L., Monreal, C.D., Miazzi, C., et al. Reproductive aspects of the Sul-Mato-Grossense native sheep. In: Annual Meeting of the Brazilian Society of Animal Science; 25–28 July. Lavras, MG: Brazilian Society of Animal Science; 2008.

[26] Sandoval Jr., P, editors. Creating manual of goats and sheeps. 1st ed. Brasília: Codevasf; 2011. 138 p.

[27] Neto, J.L. A. History of the breed: Pantaneiro sheep [Internet]. 11/01/2011. Available from: http://www.ruralcentro.com.br/analises/2214/resumo-historico-do-ovino-pantaneiro [Accessed: 08/21/2011].

[28] Morais, O.R.O. The genetic breeding of sheep in Brazil. In: Genetic breeding applied to animal production; Belo Horizonte, MG. FEPMUZ; 2001.

[29] Notter, D.R. The importance of genetic diversity in livestock populations of the future. J. Anim. Sci. 1999;77(1):61–69. DOI: 10.2527/1999.77161x.

[30] Baumung, R., Simianer, H., Hoffmann, I. Genetic diversity studies in farm animals – a survey. J. Anim. Breed. Genet. 2004;121(6):361–373. DOI: 10.1111/j. 1439-0388.2004.00479.x.

[31] Buchanan, F.C., Adams, L.J., Littlejohn, R.P., Maddox, J.F., Crawford, A.M. Determination of evolutionary relationships among sheep breeds using microsatellites. Genomics. 1994;22(2):397–403. DOI: 10.1006/geno.1994.1401.

[32] Arranz, J.J., Bayón, Y., Primitivo, F.S. Genetic relationships among Spanish sheep using microsatelites. Animal Genet. 1998;29(6):435–440. DOI: 10.1046/j. 1365-2052.1998.296371.x.

[33] Stahlberger-Saitbekova, N., Schläpfer, J., Dolf, G., Gaillard, C. Genetic relationships in Swiss sheep breeds based on microsatellite analysis. J. Anim. Breed. Genet. 2001;118(6): 379–387. DOI: 10.1046/j.1439-0388.2001.00312.x.

[34] Menezes, M.P.C., Martinez, A.M., Ribeiro, M.N., Pimenta-Filho, E.C., Bermejo, J.V.D. Genetic characterization of Brazilian native breeds of goats using 27 microsatellite markers. R. Bras. Zootec. 2006;35(4):1336–1341. DOI: 10.1590/S1516-35982006000500012.

[35] Almeida, P.A.R. Genetic diversity and differentiation of Portuguese sheep breeds based on DNA markers – microsatellites: a conservation perspective [dissertation]. Trás-os-Montes e Alto Douro University:2007. 235 p. Available from: http://hdl.handle.net/10348/118.

[36] Paiva, S.R., Silvério, V.C., Egito, A.A., MacManus, C., Faria, D.A., Mariante, A.S., et al. Genetic characterization of the Santa Inês breed. In: Second International Symposium on Sheep and Goat Production; João Pessoa, PB. State Company of Agricultural Research from Paraíba; 2003. p. 487–499.

[37] Paiva, S.R., Silvério, V.C., Egito, A.A., McManus, C.M., Faria, D.A., et al. Genetic variability of the main Brazilian hair sheep breeds using RAPD-PCR markers and conservation implications. Pesq. Agropec. Bras. 2005b;40(9):887–893.

[38] El Nahas, S.M., Hassan, A.A., Mossallam, A.A.A., Mahfouz, E.R., Bibars, M.A., Oraby, H.A.S., et al. Analysis of genetic variation in different sheep breeds using microsatellites. Afr. J. Biotechnol. 2008;7(8):1060–1068. DOI: 10.5897/AJB08.066.

[39] Paiva, S.R., Mariante, A.S., Blackburn, H.D. Combining US and Brazilian microsatellite data for a meta-analysis of sheep (Ovis aries) breed diversity: facilitating the FAO Global Plan of Action for Conservation Animal Genetic Resources. J. Hered. 2011a; 102(6):697–704. DOI: 10.1093/jhered/esr101.

[40] Ferreira, J.S.B., Paiva, S.R., Silva, E.C., McManus, C.M., Caetano, A.R., et al. Genetic diversity and population structure of different varieties of Morada Nova hair sheep from Brazil. Genet. Mol. Res. 2014; 13(2):2480–2490. DOI: 10.4238/2014.

[41] Paiva, S.R., Facó, O., Faria, D.A., Lacerda, T., Barretto, G.B., Carneiro, P.L.S., et al. Molecular and pedigree analysis applied to conservation of animal genetic resources: the case of Brazilian Somali hair sheep. Trop. Anim. Health Prod. 2011b;43(7):1449–1457. DOI: 10.1007/s11250-011-9873-6.

[42] Wood, N.J., Phua, S.H. Variation in the control region sequence of the sheep mitochondrial genome. Anim Genet. 1996;27(1):25–33. DOI: 10.1111/j.1365-2052.1996.tb01173.x.

[43] Hiendleder, S., Mainz, K., Plante, Y., Lewalski, H. Analysis of mitochondrial DNA indicates that domestic sheep are derived from two different ancestral maternal sources: no evidence for contributions from urial and argali sheep. J. Hered. 1998;89(2):113–120. DOI: 10.1093/jhered/89.2.113.

[44] Bruford, M.W., Bradley, D.G., Luikart, G. DNA markers reveal the complexity of livestock domestication. Nat. Rev. Genet. 2003;4(11):900–910. DOI: 10.1038/nrg1203.

[45] Bolzan, A.R. DNA barcoding of Micophagous drosophilids comprising the genera Hirtodrosophila, Mycodrosophila and Zigothrica [thesis]. Santa Maria, RS: Center of Natural and Exact Sciences, Federal University of Santa Maria; 2011. 85 p.

[46] Gonçalves, G.L., Moreira, G.R., Freitas, T.R., Hepp, D., Passos, D.T., Weimer, T.A. Mitochondrial and nuclear DNA analyses reveal population differentiation in Brazilian Creole sheep. Anim. Genet. 2010;41(3):308–310. DOI: 10.1111/j.1365-2052.2009.01986.x.

[47] Tserenbataa, T., Ramey, R.R., Ryder, O.A., Quinn, T.W., Reading, R.P. A population genetic comparison of argali sheep (Ovis ammon) in Mongolia using the ND5 gene of mitochondrial DNA: implications for conservation. Mol. Ecol. 2004;13(5):1333–1339. DOI: 10.1111/j.1365-294X.2004.02123.x.

Chapter 4

Livestock Methane Emission: Microbial Ecology and Mitigation Strategies

Marleny Garcia Lozano, Peña Garcia Yadira,
Karen Abigail Avendaño Arellano,
López Ortiz Carlos E. and Nagamani Balagurusamy

Additional information is available at the end of the chapter

http://dx.doi.org/10.5772/65859

Abstract

Rumen microbiome plays a critical role in the development and nutrition of the host, and any alteration in the rumen microbiome has an important effect on the animal. Rumen microbial ecology is always dynamic in response to the diets and physiological conditions of the host. Ruminal microorganisms are mainly anaerobic and provide around 75% of the energy needed by the animal. The importance of microbial diversity in rumen has gained attention not only due to its significance on the productivity of the host, but also due to the emission of greenhouse gases (GHGs) and their environmental impact. Livestock is one of the most important sources of GHGs from agriculture, contributing more than 25% of global GHGs emissions. However, the variations in livestock emission in different regions of the world could be attributed to the changes in diversity and abundance of rumen microbial communities, which vary according to the type and age of animal, type of feeds, feeding strategies, climate, etc. This chapter deals on rumen microbial ecology, the role of microorganisms in enteric fermentation and the different mitigation strategies based on manipulation of rumen microbial diversity to reduce the methane emissions from livestock.

Keywords: methane, rumen, enteric fermentation, rumen ecology, mitigation strategies

1. Introduction

Global warming has been attributed to the increment of atmospheric concentration of greenhouse gases (GHGs). Since 1750, concentrations of CO_2, CH_4 and N_2O had increased by 40, 150 and 20%, respectively, until 2014, and the rate of increment of GHG per year from 2000 to 2010 was approximately 2.2% [1]. Of various anthropogenic activities contributing to global warming, agriculture is an important source. This sector is responsible for 18% of the total anthropogenic GHG emissions annually [2]. Livestock represents the most important cause of GHG from agriculture contributing approximately to 80% of these emissions [3] and more than 25% of global GHG emissions [4].

Herrero et al. [5] estimated the total emissions from livestock were in the range of 5.6 – 7.5 $GtCO_2$–eq/year (5.6 to 7.5 × 10^{12} kg CO_2–eq) between 1995 and 2005. They observed that the main sources were enteric CH_4 (~32.2%), N_2O emissions associated with feed production (~27.45%) and land use for animal feed and pasture (~24.42%). Havlík et al. [6] opined that ruminants represent more than 80% livestock emissions; particularly, beef and dairy sector contribute to about 60% [7]. Emissions from enteric fermentation contribute to 8% of total CH_4 emissions and are estimated to increase to 30% between 2000 and 2020 [8]. Enteric fermentation is the normal process of feed digestion in ruminants and is mediated by the microbial activity in the rumen and in the large intestines. Significant amount of methane is produced by methanogens residing within the rumen (87%) [9], which is released principally through eructation, approximately 10–15% is emitted by normal respiration and via flatus [10].

The continued growth of human population and consequent demand for food are potential drivers of GHG emissions. International climate negotiators have been focused to reduce GHG emissions by the improvement of engineering processes, energy efficiency and investments on alternative energy generation technologies. However, the abatement of ruminant GHG emissions has not received adequate attention by the United Nations Framework Convention on Climate Change [11]. Even so, several research groups have been working to develop strategies to optimize ruminal functions in order to achieve the desired levels of production by enhancing feed conversion efficiency and simultaneously reducing methane emissions by manipulating the rumen microorganisms. It is essential to have a detailed knowledge of ruminal microbiome, their interactions among themselves and with the host to achieve these objectives, and to identify the new approaches for mitigation of GHGs emissions [12].

2. Livestock GHG emissions

Livestock emissions depend considerably on some of the environmental characteristics such as the mean annual temperature, geographic location and the economic level of the country. It has been observed that in developing and emerging countries, the dietary habits increase meat consumption contributing to these emissions [4, 13], nevertheless developed countries have a greater proportion of intensive animal production, which results in higher emissions of CH_4, which is estimated to be 150.7 g/cow/day by cattle [4]. Additionally, the size and productivity of animals affect their feed intake and enteric CH_4 emissions [14], which can vary by animal type, growth stage and composition of diet [15, 16]. Castelán-Ortega et al. [17] reported that the average CH_4 emissions by individual dairy cattle are higher in the tropics than in temperate regions, 319.1 and 283 g/day, respectively. This could be attributed to the elevated proportion of cellulose in tropical forages, which is reported to produce three times more CH_4 than hemicellulose.

http://dx.doi.org/10.5772/65859

The estimation of livestock emissions differs considerably between studies as different models are employed for their estimation. Some authors use their own models, but most of the authors follow the guidelines of IPCC [18]. However, the differences on estimations still continue. Tier I utilizes default global or regional emission factors. Tier II utilizes estimated regional or local emission factors and is used in some enteric fermentation studies, nevertheless Tier III is the most reliable model for enteric CH_4 emission and has several advantages compared to Tier II, because it represents mechanisms of enteric fermentation in more detail and can be expected to describe more of the variations caused due to nutritional and animal factors [8, 19].

Enteric fermentation in ruminants and manure management emissions contributes directly to around 9% of total anthropogenic emissions. In 1990, enteric methane global emissions were 84 Tg/year CO_2-eq (84×10^9 kg CO_2-eq), which increased to 92 Tg/year CO_2-eq in 2005. It is reported that the main sources of global enteric CH_4 emissions are Asia (33%), followed by Latin America (23.9%), Africa (14.5%), Western Europe (8.3%) and North America (7.1%) [14]. Beef trades also have a significant impact on GHG emissions. Emissions from beef trade represented 2% of total emissions traded internationally in 2010 and increased by 19% during the period between 1990 and 2010. The dominant global fluxes in 2010 were the exportation of emissions embodied in meat from Brazil and Argentina to Russia (2.8 and 1.4 Mt CO_2-eq (2.8 and 1.4×10^9 kg CO_2-eq), respectively), emissions embodied in US imports of meat from Canada were the same that emissions embodied in US exports to Mexico of 1.2 Mt CO_2-eq. Australian meat exported to South Korea also embodied substantial emissions of 1.0 Mt CO_2-eq. In European countries, meat exported from France to Italy and France to Greece embodied 1.4 and 1.2 Mt CO_2-eq emissions, respectively. Also Italian meat imported from Poland, Germany and Netherlands embodied 0.7, 0.6 and 0.7 Mt CO_2-eq emissions, while Chinese emissions embodied in beef exported were small in comparison with the other countries. Although emissions due to import of meat are considered insignificant, it is important to consider all livestock sectors that contribute to emissions [13].

With respect to the Mexico, total CH_4 emissions in 2006 were 8954.10 Gg, and agriculture sector was the highest contributor with significant input due to enteric fermentation and manure management [16]. Earlier, Rendón-Huerta et al. [18] has also reported that enteric CH_4 emissions are the major source of GHG emissions in Mexican livestock production systems. They calculated the GHG emissions from dairy cattle in Mexico for a period of time of 30 years using a Tier II of IPCC and reported that emissions of CH_4, N_2O and CO_2-eq during 1970 to 2010 increased from 144 to 270, 0.349 to 0.713 and 3704 to 6962 Mt/year, respectively. They observed that methane emissions per cow increased by 11%, while per liter of milk decreased by 30%. In the past 40 years, total N_2O emission increased by 104%, but N_2O/cow emissions increased only by 22% in the same period and decreased by 25% per liter of milk. The reduction in GHG emissions per liter of milk means an increase in the efficiency of production systems resulting in an augmentation of milk production per cow and consequentially diminishing the emissions [18]. Hernández-De Lira et al. [16] based on animal census data from 2012, reported that the methane emissions by enteric fermentation in Mexico were 1926.08 Gg CH_4, of which beef cattle produced 1651.8 Gg CH_4; while dairy cows generated only 172.70 Gg CH_4.

Emissions by manure management, mostly CH_4 and N_2O, are produced during the manure decomposition carried out by anaerobic microbial activities. These emissions depend on specific manure composition and quantity produced which, in turn is dependent on other factors as animal type, breed, weight, diet and climate conditions. Although CH_4 emissions from enteric fermentation are higher than those from manure [13, 16], manures also contribute to N_2O emissions due to volatile nitrogen losses, principally in form of ammonia (NH_3) and NO_x [13]. They have reported that CH_4 and N_2O emissions from manure would increase by 20 and 29%, respectively, from 2000 to 2020.

Asia, particularly China, Western Europe and North America are the regions with the highest GHG emissions from manure management [14]. According to EPA [20], global GHG emissions from manure management were 446 million tonnes of CO_2-eq, of which the share of CH_4 and N_2O was 53 and 47%, respectively, while FAO [3] estimated global GHG emissions from manure management were 368 million tonnes of CO_2-eq. In case of Mexico, CH_4 emission from manure was 62.24 Gg CH_4, where beef cattle and dairy cow emitted with 29.49 and 2.42 Gg CH_4, respectively [16]. Similarly, FAO [3] reported that Asia, Central and South America, Sub-Saharan Africa, Western Europe, North America, Eastern Europe and the Commonwealth of Independent States were the regions with the highest emissions of N_2O due to manure [14].

3. Rumen environment

Ruminants are herbivorous mammals considered as latecomers in evolution. Their forestomach is a very complex environment, which allows them to convert plant tissues into nutritious and useful products. The digestive tract of ruminants is formed by various compartments such as reticulum, rumen, omasum, abomasum, small intestine, cecum, colon and rectum [21]. The ruminant stomach is composed by three pregastric fermentation chambers (rumen, reticulum and omasum) [22] (**Figure 1**). Environmental conditions such as temperature (38–42°C), redox potential (250 to 450 mV), pH (5.5–7) controlled by buffer in saliva and osmolarity (260–340 mOsm) [23] provide the ideal conditions for the digestion of plant material by microorganisms. Fibrous components are hydrolyzed and fermented by the interactions among different microbial communities inhabiting the rumen, producing mainly acetate, propionate and butyrate, CO_2, H_2 and CH_4. VFAs are the most important source of energy for the animal (75% of the total amount of the digested energy) [24]. Moreover, microbial cell biomass is the major source of protein and amino acids [25]. Microbial population also synthetizes vitamins B and K and employs detoxification mechanisms for phytotoxins and mycotoxins [26].

Microbial ruminant ecosystem is composed by a high microbial population density, predominantly obligate anaerobic microorganisms. Bacteria are the most abundant microorganisms and more than 50% of the cell mass in the rumen are comprised of at least 50 bacterial genera (10^{10} –10^{11} ml^{-1}), followed by 25 genera of ciliate protozoa (10^4 –10^6 ml^{-1}), six genera of fungi (10^3–10^6 ml^1), methanogenic archaea (107–1010 ml^1) and bacteriophages (108–109 ml^1) [27–29], nevertheless only 10% of these microbiome have been identified and described [30].

http://dx.doi.org/10.5772/65859

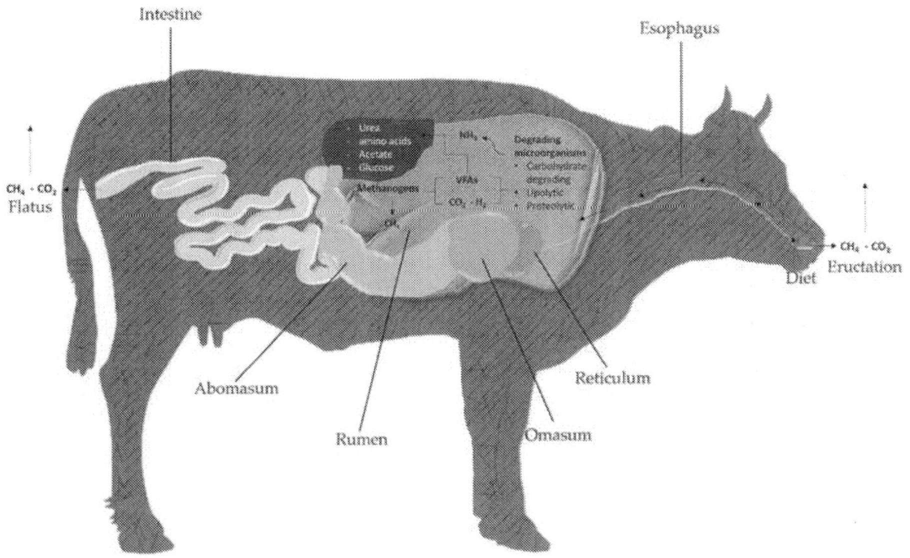

Figure 1. Ruminant digestion process. Note: Gastrointestinal tract of ruminants and main biochemical processes occurr ing in it.

The interactions of these microorganisms are widely different, namely mutualism, commensalism, syntrophy, competition and depredation [31, 32].

Hydrolysis of plant polysaccharide material is the first step in the enteric fermentation process, and 80% of plant cell material degradation is carried out by bacteria and fungi, and the rest 20% is by protozoa [33]. In the second stage, monomers are fermented to VFAs, branched chain VFAs, organic acids (lactate), alcohols, CO_2 and H_2. VFAs are absorbed by the rumen and omasal walls of the host animal for its nutrition [10]. Though several parameters such as rumen fluid, volume, pH and VFAs, concentration can disturb this absorption [34]. Free acids can be oxidized by obligate hydrogen producing bacteria to acetate, albeit this reaction is thermodynamically non-favorable, and hence are carried out only in synthropic association with hydrogen consuming bacteria or archaea, which diminish the partial pressure of H_2. When the conditions are not favorable, VFAs are accumulated, decreasing the pH and inhibiting rumen microbiome [35, 36]. NH_3 is produced due to proteolysis and can be used by microorganisms to build their own proteins. The excess of NH_3 is absorbed by the rumen wall and transported by the animal blood [37]. The digested proteins, lipids and the carbohydrate constituents of microbial cells are exploited in the small intestine for the maintenance of the animal and the production of meat and milk. During enteric fermentation, a large quantity of CO_2 is produced due to diverse biochemical processes. A part of this CO_2 produced is released through eructation or normal respiration, and other part is reduced with H_2 to CH_4 by hydrogenotrophic methanogens. Methane produced is primarily released through eructation and approximately 10–15% is emitted by normal respiration and via flatus [10].

CH_4 production can be accomplished by the reduction of acetate and methyl-containing C1 compounds, nonetheless these pathways are not common in the rumen [38]. About 2–12% of gross energy intake (GEI) produced in the rumen by fermentation is converted to methane, which apart from leading to the loss of the feed energy, results in the emission and consequently, global warming [39].

4. Microbial diversity and abundance in rumen

As explained above, microorganisms present in gastrointestinal tracts (GIT) of ruminants and their relationship yield several benefits to the host. The composition of microbiome in GIT varies according to several conditions. Microbial populations can be affected by factors such as type and race of animal, age of the host, diets, feeds, farming practicing and geographical regions [40].

The microbial diversity presents in ruminant's changes across different points of the GIT. Mao et al. [41] studied the microbial population of 10 distinct sites of the GIT in dairy cattle and observed that the microbial diversity differed for the analyzed points. They reported 21 different phyla belonged to Firmicutes (64.81%), Bacteroidetes (15.06%) and Proteobacteria (13.29%). At genus level, the most abundant genera in cattle GIT included *Prevotella, Treponema, Succiniclasticum, Ruminococcus, Acetitomaculum, Mogibacterium, Butyrivibrio* and *Acinetobacter* as well as many different unclassified genera, among which Prevotella, unclassified Ruminococcaceae, unclassified Rikenellaceae, unclassified Christensenellaceae and unclassified Bacteriodales were predominant.

A study carried out by Henderson et al. [42] determined the rumen microbiology of 32 species or subspecies of animals from 35 different countries of seven world regions and evaluated the differences among them. Seven bacterial groups comprised around 67.1% of the total bacterial sequenced, they corresponded to *Prevotella, Butyrivibrio* and *Ruminococcus,* as well as unclassified *Lachnospiraceae, Ruminococcaceae, Bacteroidales* and *Clostridiales,* but were not present in the same proportions in all animal species tested. The abundance of archaea worldwide was similar in all the sampled analyzed, and all belonged to methanogens and corresponded to *Methanobrevibacter gottschalkii* and *M. ruminantium. Methanosphaera* sp. and two *Methanomassiliicoccaceae*-affiliated groups, contributing to 89.2% of total archaeal community in rumen. Even in the same region, the age of the animal is other important factor that contributed to considerable differences in microbial diversity. It has been demonstrated that the ruminal microbiota of young dairy cattle is more heterogeneous than microbial community of those cows reaching maturity (2 years). In general, microbial communities in the rumen of dairy cows have been dominated by bacteria (>90%), followed by eukarya (2–8%) and a small abundance of archaea (1.0%). Similarly, a metagenomic study of the rumen microbiome in Holstein dairy cows reported 26 bacterial phyla belonging to Bacteroidetes (61–80%), followed by Firmicutes (12–23%), Proteobacteria (3–10%), Spirochaeta, Fibrobacteres and Actinobacteria (up to 2%). Again, they reported that *Prevotella* from Bacteroidetes was the most abundant genus (>50%), followed by *Bacteroides* (10.91%) and *Parabacteroides* (1.73%). In the case of Firmicutes, the predominant genera were *Abiotrophia, Acetivibrio* and *Acetohalobium.*

In the archaeal community, the genera *Methanobrevibacter,* being the predominant genera, and accounted 0.5% of the total microbial abundance [43].

Earlier, Kim et al. [44] analyzed the diversity of bacteria and archaea based on 16S ribosomal RNA (rRNA) and reported 13,478 bacterial and 3516 archaeal sequences, which correspond to 7000 and 1500 species of bacteria and archaea, respectively. Among nineteen phyla of bacterial domain, the most abundant were *Firmicutes* (57.9%), *Bacteroidetes* (26.7%) and *Proteobacteria* (6.9%). Within *Firmicutes,* the most abundant class was *Clostridia* (>90%), and the rest belonged to *Bacilli, Erysipelotrichi* and unclassified *Firmicutes*. In the *Clostridia* class, the predominant genera were *Buryrivibrio, Acetivibrio, Ruminococcus, Succiniclasticum, Pseudobutyrivibrio* and *Mogibacterium*. In the *Bacteroidetes* phylum, the predominant class was *Bacteroidia,* and *Prevotella* represented the most abundant genera. All the five classes of *Proteobacteria* were represented in the rumen bacterial sequences. More than 99% of the archaeal sequences correspond to the phylum *Euryarchaeota,* followed by 11 sequences of the phylum *Crenarchaeota*. About 94% of all archaeal sequences were assigned to the classes *Methanobacteria, Methanomicrobia, Thermoplasmata* and *Methanopyri,* all of them within phylum *Euryarchaeota*. However, this microbial abundance in rumen can be considerably different between the extremely high and low methane emitters. While archaea are 2.49 times more, bacteria are less (0.98×) in high emitters. In addition, *Euryarcheota* and *Crenarcheota* recorded an increase in high emitters (2.48× and 3.00×, respectively), and at genus level, *Methanobrevibacter* and *Methanosphaera* have been found more abundant (2.44× and 2.54×, respectively). In case of bacterial domain, there were no significant differences between *Firmicutes* and *Bacteroides* between high and low emitters, but *Proteobacteria* was 0.24 times less in high emitters. At genus level, *Desulfovibrio* was two times more in high emitters than low emitters. However, a higher abundance of *Succinovibrionaceae* was recorded in low emitters along with a change in acetate and hydrogen concentration profile, resulting in a low methanogenesis [45]. These microbial dynamics in animals of different types and from different regions clearly demonstrate that it is possible to develop strategies to mitigate livestock methane emission through microbial manipulation strategies. Various studies [46, 47] have suggested that it is possible to adapt the rumen microorganisms by manipulating the feeding management in the young animal, which have been found to persist in their later life. These results suggested that the methane emissions can be decreased considerably by manipulation of rumen microbiome through feed alterations.

As mentioned earlier, the composition of population in rumen is affected by the age and diet of the animal. Li et al. [47] evaluated the rumen microbiota of pre-ruminant calves of 14- and 42-day-old calves fed milk replacers based on 454-pyrosequencing of 16S rDNA and reported a total of 170 bacterial genera in the developing rumen of 14-day-old calves. They, further demonstrated that microbiota changed according to their dietary modifications and physiological changes in the host. Moreover, the transition from 14 to 42 days had a significant impact on the ruminal microbial composition. The most abundant phylum, *Bacteroidetes,* increased significantly his abundance from 45.7 (14 days) to 74.8% (42 days), the phylum *Synergistetes* also increased, while the abundance of *Firmicutes, Proteobacteria* and *Fusobacteria* decreased during this time. The results of these two age groups are different from those based on the rumen of 12-month-old animal, where the most abundant phyla were *Bacteroidetes* (52%), *Firmicutes* (42.7%), *Spirochaetes* (2.3%) and *Fibrobacteres* (1.9%). This study clearly demonstrated that the changes in feed affect and change the dynamics of ruminal microbiome. Petri et al. [48] studied the impact of diet and its impact of an acidotic challenge on the composition of six different bacterial targets from heifers fed forage, mixed forage, high grain, post-acidic challenge (4 and 12 h) and recovery. They observed that all of the bacterial target groups were affected by dietary treatment, with exception of *S. bovis*, *Raminococcus* spp. and *Fibrobacter succinogenes* represented a large percentage of the bacterial population present in the mixed forage diet. *Prevotella* corresponds to the most abundant genera in the acidotic challenge, but the lowest in the animal fed forage. *Megasphaera elsdenii* was present in abundance in the sample of 12 h after acidotic challenge, but its abundance decreased during recovery, while at the same time *S. ruminantium* increased in proportion. Both *S. ruminantium* and *M. elsdenii* accounted the smallest proportion of the bacterial population in heifers fed forages.

5. Methane mitigation strategies

The necessity to implement abatement strategies for enteric GHG emissions has been expanded in conjunction with the increase in the population and food demand. There are two concerns over methane emissions by livestock ruminants. First, the release of methane is considered a loss of energy for the animal, resulting in a decrease in animal productivity between 2 and 12%. Second, the calorific potential of methane released has a negative impact on climate change. There are several publications on strategies to reduce methane production [49–52]. The main target of these strategies is on methanogenic archaea by decreasing their substrate availability either directly or indirectly. Overall, abatement strategies include mechanisms such as modifications in dietary composition, and/or by supplementation of diet with chemical inhibitors, lipids or plant compounds, some of these strategies are shown in **Figure 2**.

5.1. Dietary composition

The quantity of enteric methane production is directly related to the quantity and quality of the feed consumed by the animal. The loss of GEI was augmented with an increase in high feed quantity. Animals with a low feed efficiency increase environmental impact due to the loss of GEI in form of methane.

The most common feeding mechanisms for the ruminants are based on pasture (grazing) and harvested forages. Hay and silage are the most common cattle forages. Hay has been recognized as superior feed than silage, but in cold and wet weather, silage is most used due to its major productivity. Silages for ruminants in temperate areas are usually based on cereals and legumes such as grass, maize, lucerne and red clover, which provide carbohydrate, protein and lipid sources for the animal [53]. It has been extensively reviewed that the replacement of ruminant forage diets with high grain diets can reduce methane production [9, 27, 54, 55]. Fermentation of cereal grains with high starch content increased the voluntary intake and reduced the residence time in the rumen, promoting post-ruminal digestion. Starch also

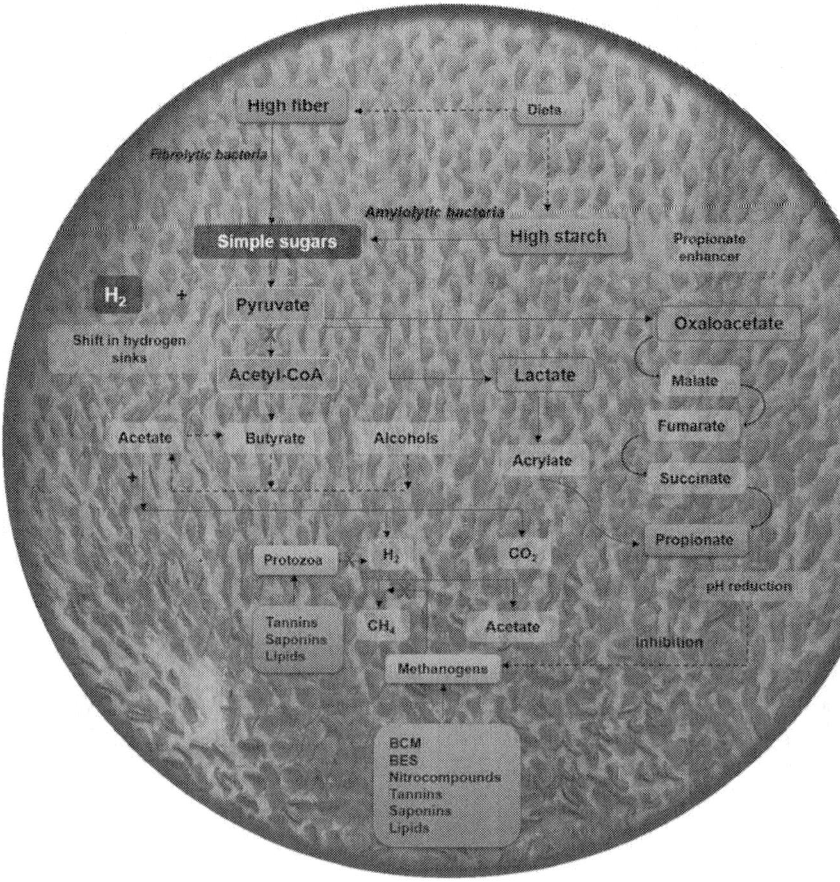

Figure 2. Mitigation strategies on methane emission by rumen microbiome manipulation through change in diets. Note: The main pathways and products formed when high fiber diet is used are represented in green color. The effect of high starch diets, which enhances propionate production due to shifting of hydrogen sinks, is presented in orange color. Dietary supplements and their main targets in order to reduce methane production are indicated in blue color.

enhanced propionate production, which depleted H^+, and thereby decreased its availability for hydrogenotrophic methanogens. Moreover, propionate production decreased the pH, causing an inhibitory effect on methanogens and protozoa [56]. The loss of GEI with grain-based diets is commonly 4%, while it is 6.5% or more in forage-based diets.

Lettat et al. [57] reported that starchy diets, apart from increasing the propionate concentration, decreased the concentrations of acetate and butyrate and consequently methane production (-14%). Diversity and richness of bacterial community were reduced with increase in the starch content of the diet, however, the total bacterial population, *Prevotella* spp. and *M. elsdenii* were favored. The bacterial group *Prevotella* has been identified as amylolytic and propionate producer and the dominant within the rumen [58, 59], while *M. elsdenni* is a well-known lactate-utilizing and propionate-producing bacteria. CH_4 reduction has been linked to the decrease in protozoan populations since protozoa are known as hydrogen producers and are in symbiotic relationship with methanogens. Hence, with a decrease in protozoan population, there is a decrease in the hydrogen transfer between them and methanogens, and this decreased the methane production. However, metabolic activity of archaea and methanogenic population increased when methane production decayed, demonstrating the cDNA-qPCR method to estimate archaeal growth and activity is unreliable to reflect changes in ruminal methanogenesis. However, there should be sufficient care before adopting this as a wholescale strategy. It has reported that changes in dietary composition not only can affect microbial diversity but also can generate animal disorders, producing a negative effect on the host. Saleem et al. [60] reported that high grain diets increased the concentrations of several toxic compounds such as putrescine, methylamines, ethanolamine and VFAs in the rumen fluid. VFAs accumulation can decrease the pH lower than 5.5 and produce subacute ruminal acidosis, which is a common and disturbing problem for farmers [61]. High grain diets have been commonly observed in favor of amylolytic microorganisms and against fibrolytic microorganisms. Petri et al. [48] reported that rumen of Angus heifers fed with high grains diet recorded a higher abundance of *Prevotella* spp., *S. ruminanitum* known also as amylolytic bacteria, and *M. elsdenii*. Whereas, a higher abundance of the fibrolytic bacteria *Ruminococcus* spp. and *F. succionogenes*, and the lactate-producing *S. bovis* was observed with forage diet. Kittelmann et al. [62] observed a positive correlation between the occurrence of methanogens and fibrolytic bacteria. *Methanobrevibacter ruminanitum* is found to be correlated with the family Fibrobactereacea and *M. gottschalkii* with the family *Ruminococcaceae*. *Ruminococcus* spp. is known to produce large amounts of H_2, while *Fibrobacter* spp. produces formate, which is substrates for methanogens. Therefore, the abundance of fibrolytic bacteria could be related with methanogenic communities and consequently with methane production.

5.2. Dietary supplementation

5.2.1. Chemical inhibitors

Compounds nontoxic to animal, but inhibitors to methanogens have been used to reduce methane production. Although these compounds inhibit-specific enzymes involved in methanogenesis pathway, it has been reported that they could also have an impact on other microbial groups present and could affect the uptake of feed by the animal [5, 27]. The most used and effective compounds are the analogous of coenzyme M, inhibitors of methanopterin biosynthesis, nitrocompounds and halogenated compounds [63–65].

Bromochloromethane (BCM), a methane analogue, has been extensively used to decrease methane production [65–67] but has a limited use due to its great ozone depleting capacity [66]. This compound reduces vitamin B12 and inhibits the cobamide-dependent methyl transferase step of the biosynthesis pathway of methyl coenzyme M, involved in methanogenesis pathway. After 12 h of supplementation, BCM-cyclodextrin (0.5 g/100 kg live weight) decreased the methane production of steer by 29%, and without adversely affecting the animal productivity [65]. Mitsumori et al. [67] studied the effect of different concentrations of BCM-cyclodextrin (BCM-CD) on the rumen microbial population of goats. Doses of BCM-CD were of low (0.5 g/100 kg live weight LW), medium (2 g/100 kg LW) and high (5 g/100 kg LW), which decreased the methane emissions by 4.64, 71.46 and 91.23%, respectively. Denman et al. [68] analyzed the microbial diversity of the samples from the above study and reported that the relative abundance of Bacteroidetes increased with the BCM-CD doses, while Firmicutes, Synergistetes and Lentrisphaerae phyla decreased. In the case of control animal, Bacteroidetes (60%) was dominant, followed by Firmicutes (24%), Synergistetes and Lentisphaera (both contributed ~4%). Administration of BCM also reduced considerably methanogenic diversity, however, *Methanobrevibater* species were the most abundant in all treatments. Based on phylogenetic binding and functional assignment, the major genera were *Prevotella* and *Selenomonas* which were associated with the propionate production by the randomizing succinate pathway. This pathway was the primary route of H_2 consumption and decreased H_2 availability for methanogens.

2-bromoethanesulfonate (BES) is another common and successful compound to decrease methane emissions, which is an analog of coenzyme M. In an *in vitro* mesocosm study with cow manure and anaerobic digester sludge, a 89 and 100% decrease in methane production was observed at 0.5 and 10 mmol/L, respectively. Relative abundance of *Methanosaeta* and *Methanosarcina* decreased considerably at 10 mmol/L. Moreover, a decrease in mcrA expression, which encodes the α subunit of the methyl coenzyme M reductase and due to it is used for the relative measure of methane metabolites and methanogenic abundance in different environments [69], was observed with the increment of BES. A decrease in syntrophic-bacteria *Syntrophomonas* was observed too at both concentrations of BES. It is known for oxidation of butyrate and other fatty acids in syntrophic association with H_2-consuming bacteria and/or hydrogenotrophic methanogens and could explain the decrease in methanogenic activity [70].

The inhibitory effect of chloroform is attributed to its capacity to target the corrinoid-containing MtrA subunit of the large multimeric membrane enzyme methyl tetrahydromethanopterin:coenzyme M methyltransferase [71]. Martínez-Fernández et al. [72] studied the inhibitory effect of chloroform-cyclodextrin (CCD) by way of supplementation; as low (1 g/100kg live weight LW), medium (1.6 g/100 kg LW) and high (2.6 g/100 kg LW) dose along with two diets (roughage:concentrate (60:40) or roughage hay) in eight steers. All three doses decreased the methane production by 14, 37 and 55%, respectively. Changes in microbial community were observed too, archaeal abundance was negatively correlated with CDD levels, Methanobacteriaceae family and Methanoplasmatales order were found to be decreased. Protozoan population increased with CCD doses with roughage:concentrate diet, while chloroform did not have any effect on fungi community. Bacterial population was also affected, relative abundance of *Bacteroidetes* increased, while *Firmicutes, Synergistetes* and *Verrucomicrobia* phyla were decreased. While methanogenesis was inhibited, an increment in the production of amino acids, organic and nucleic acids was observed. All of these metabolic changes modified the ruminal microbiome, increased the *Bacteroidetes:Firmicutes* ratio and decreased archaea and *Synergistetes*. Although abundance of fibrolytic bacteria, protozoa and fungi was not affected, methanogenesis was inhibited by 30%. They concluded that the use of chloroform as methanogenic inhibitor did not adversely affect rumen metabolism and could redirect H_2 to another pathways producing non-methane end products.

Apart from the compounds mentioned, nitrocompounds are also being used in vivo to mitigate methane emissions. These compounds target of specific sites of MCR due to its molecular shape and oxidative potential and inhibit the last step of methanogenesis pathway. It has been reported that 3-nitrooxypropanol (NOP) at 40–80 mg/kg, decreased methane emissions around 30% and also increased body weight gain considerably without affecting feed intake or milk characteristics [73]. Duin et al. [74] reported that only 0.1 μM NOP is needed to inactivate completely MCR, and 1 μM to inhibit the methanogenic population. It was also reported that bacterial population was not affected by the addition of NOP, while methanogenic population decreased and protozoal abundance increased [75]. The decrease in methane production (−59.2%) by NOP (2 g/day) could be related directly to the reduction in the population of methanogens. The reduction in methanogen populations due to the addition of nitrocompounds need not always result in an increase in protozoan populations, since the compounds could also affect the symbiotic methanogens-protozoan association and thereby could result in decreased protozoan populations.

5.2.2. *Plant bioactive compounds*

Plant secondary metabolites have also been extensively used in the reduction of methane emissions. The most common used are tannins, saponins and essential oils, and they can affect methanogens either directly or indirectly. Further, they reduce protozoal population and thereby reducing symbiotically associated methanogens, apart from decreasing fiber digestibility and H_2 production [76].

Tannins are polyphenolic compounds which form complexes with metal ions, amino acids and polysaccharides, and thereby reduce ruminal fermentation. They can be divided into hydrolysable and condensed tannins. Hydrolysable tannins at high concentrations may be toxic to ruminants, while condensed tannins can make several nutrients unavailable to the animal due to irreversible binding [77]. Moreover, they can bind to the gastrointestinal tract, causing negative effects [78]. However, they have been found to be effective in reducing methane emissions. Condensed tannins have been reported to reduce methane by around 16% based on dry matter intake (DMI) [79]. Total methanogen population decreased by 22.3–36.7% when purified hydrolysable (HT) and condensed tannins (CT) (1 mg/ml) were tested in vitro conditions. Hydrolysable tannins were found to be more effective than with condensed tannins in reducing methane formation [80]. On the contrary, Bhatta et al. [76] reported that CT had a greater effect on methane reduction (−5.5%) than HT (−0.6%) and its inhibitory effect on methanogens (−28.6%) was more than HT (-11.6%). Protozoan populations also decreased by 12.3% with HT diets. However, a combination of HT+CT diets had a more significant effect and a 36.2% decrease was reported. Although tannins reduced total VFA concentrations was found to increase propionate concentrations and decrease iso-acids, which could have a negative effect on methanogenesis. In previous studies, a reduction in total and cellulolytic bacteria in response to tannins was observed along with the reduction in VFA production and also H_2 production, contributing to methane inhibition [81, 82].

Saponins are complex and diverse molecules which are divided in triterpene and steroid glycosides [83]. They are considered effective compounds to suppress methane production due to their anti-protozoan properties [54]. Their anti-protozoan properties are attributed mainly to the formation of complexes with sterols in the membrane surface of protozoans [84]. However, this is pH dependent and composition of diet with addition of saponins [85, 86]. Moreover, saponins are potential defaunation agents and could result in the reduction in enteric CH_4 production by eliminating protozoa [9]. Nevertheless, they have an effect on the whole ruminant microbiome and animal digestion process, and not specifically targeting protozoan populations.

5.2.3. Lipid supplementation

Supplementation of lipids in ruminant diets is found to improve microbial metabolism of rumen, decreasing enteric methane emissions. Reduction in methane production could be due to the direct effect of fatty acids on methanogens, or indirectly due to the inhibition of the protozoan communities and associated methanogens due to enhanced propionate production. Beauchemi et al. [54] calculated that CH_4 (g/kg DMI) is reduced by 5.6% for each percentage unit of lipid, while Eugène et al. [87] estimated the methane reduction to about 2.3%.

Lipids commonly supplemented to reduce enteric fermentation are calcium salts of fatty acids, hydrogenated fats, and fats of animal origin, extracted plant oils, oilseeds and wastes from processing plants with high fat content [88]. Based on a meta-analysis of 27 publications on the effect of fatty acids in ruminant diets, fatty acids C12:0 and C18:3 demonstrated a significant inhibitory effect on methanogenesis without affect the productivity in dairy cattle [89]. Patra and Yu [90] analyzed *in vitro* the effect of five essential oils (EO) such as clove oil (CLO; from Eugenia spp.), eucalyptus oil (EUO; from *Eucalyptus globulus*), garlic oil (GAO; from *Allium sativum* L.), origanum oil (ORO; from *Thymus capitatus* L. Hoffmanns & Link) and peppermint oil (PEO; from *Mentha piperita* L.) on methane production, fermentation and ruminal microbiome. CLO, EUO, GAO, ORO and PEO significantly reduced the methane formation by 34.4, 17.6, 42.3, 87 and 25.7%. Further, decrease in relative abundance of ruminant microbial population such as archaea, protozoa and major cellulolytic bacteria *F. succinogenes*, *R. flavefaciens* and *R. albus* was recorded. Microarray analysis by RumenBactArray showed that the effect of each oil tested was unique. *Firmicutes* phylum was decreased by addition ORO and GAO, but increased by PEO. While, *Bacteroidetes* phylum, mainly *Prevotella* OTUS were found to be increased by addition of ORO and PEO. EO decreased the abundance of several microorganisms, *Syntrophococcus sucromutans*, *Succiniclasticum ruminis* and *Lachnobacterium* and members of *Lachnospiraceae*, *Ruminococcaceae*, *Prevotellaceae*, *Bacteroidales* and *Clostridiales*. This was correlated with feed degradability, ammonia concentration and molar percentage of VFAs, which directly affect microbial communities, their metabolic interactions and hence the methane production.

Beauchemi et al. [91] studied the effect of addition of saturated and unsaturated long-chain fatty acids to cattle basal diet, consisting mainly of whole-crop silage. Lipids of animal origin (tallow) and sunflower oil at 34 g/kg, and oilseed (whole sunflower seeds) at 89.3 g/kg were added to bring the total dietary fat content to about 59 g/kg of dry matter. On basis of dry matter intake, diets containing tallow or sunflower oil decreased methane emissions by 11%, while sunflower seeds by 23%. Based on digestible energy intake, all lipid sources decreased methane emissions by 17%. Previously, coconut oil has also been reported as an effective inhibitor of methane production. Jordan et al. [92] reported a 39% decrease in methane emission at a concentration of 375 g/day.

Although supplements are being used primarily in reducing methane emission from livestock, their use in increasing efficiency in feed conversion and animal productivity, based on GEI, animal weight gain, meat and milk production has also been reported [73]. However, few other studies also have reported the negative effect of supplements on the quantity and quality of animal products such as milk and meat [60, 61]. This contradiction could be due to the reason that rumen microbial diversity is dependent on type and amount of feed, which in turn influences the nutrient absorption by animal. This implies that further studies on the relation between rumen microbiome and metabolomics of rumen are essential in order to understand the variations in relation to animal products due to supplements.

Author details

Marleny Garcia Lozano, Yadira Peña Garcia, Karen Abigail Avendaño Arellano, Carlos E. López Ortiz and Nagamani Balagurusamy*

*Address all correspondence to: bnagamani@uadec.edu.mx

Laboratorio de Bioremediación, Facultad de Ciencias Biológicas, Universidad Autónoma de Coahuila, México

References

[1] IPCC (Intergovernmental Panel on Climate Change). Climate Change 2014: Mitigation of Climate Change. Cambridge University Press, New York, USA. 2015, p. 3.

[2] Steinfeld H, Gerber P, Wassenaar TD, Castel V, de Haan C. Livestock's Long Shadow: Environmental Issues and Options. Food Agriculture Organization, Rome, Italy. 2006.

[3] FAO. Food and Agriculture Organization of the United Nations Rome, Italy. FAOSTAT. 2006.

[4] Sejian V, Lal R, Lakritz J, Ezeji T. Measurement and prediction of enteric methane emission. International Journal of Biometeorology. 2011;**55**(1):1–16. doi:10.1007/s00484-010-0356-7

[5] Herrero M, Havlík P, Valin H, Notenbaert A, Rufino MC, Thornton PK, Blümmel M, Weiss F, Grace D, Obersteiner M. Biomass use, production, feed efficiencies, and greenhouse gas emissions from global livestock systems. Proceedings of the National Academy of Sciences. 2013;**110**(52):20888–20893. doi:10.1073/pnas.1308149110

[6] Havlík P, Valin H, Herrero M, Obersteiner M, Schmid E, Rufino MC, Mosnier A, Thornton PK, Böttcher H, Conant RT, Frank S, Fritz S, Fuss S, Kraxner F, Notenbaert A. Climate change mitigation through livestock system transitions. Proceedings of the National Academy of Sciences. 2014;**111**(10):3709–3714. doi:10.1073/pnas.1308044111

[7] Bellarby J, Tirado R, Leip A, Weiss F, Lesschen JP, Smith P. Livestock greenhouse gas emissions and mitigation potential in Europe. Global Change Biology. 2013;**19**(1):3–18. doi:10.1111/j.1365-2486.2012.02786.x

[8] Bannink A, Van Schijndel MW, Dijkstra J. A model of enteric fermentation in dairy cows to estimate methane emission for the Dutch National Inventory Report using the IPCC Tier 3 approach. Animal Feed Science and Technology. 2011;**166–167**:603–618. doi:10.1016/j.anifeedsci.2011.04.043

[9] Boadi D, Benchaar C, Chiquette J, Massé D. Mitigation strategies to reduce enteric methane emissions from dairy cows: Update review. Canadian Journal of Animal Science. 2004;**84**(3):319–335. doi:10.4141/A03-109

[10] Hill J, McSweeney C, Wright ADG, Bishop-Hurley G, Kalantar-Zadeh K. Measuring methane production from ruminants. Trends in Biotechnology. 2016;**34**(1):26–35. doi:10.1016/j.tibtech.2015.10.004

[11] Ripple WJ, Smith P, Haberl H, Montzka SA, McAlpine C, Boucher DH. Ruminants, climate change and climate policy. Nature Climate Change. 2014;**4**(1):2–5. doi:10.1038/nclimate2081

[12] Chaucheyras-Durand F, Ossa F. Review: The rumen microbiome: Composition, abundance, diversity, and new investigative tools. The Professional Animal Scientist. 2014;**30**(1):1–12. doi:10.15232/S1080-7446(15)30076-0

[13] Caro D, LoPresti A, Davis SJ, Bastianoni S, Caldeira K. CH4 and N2O emissions embodied in international trade of meat. Environmental Research Letters. 2014;**9**(11):114005–114018. doi:10.1088/1748-9326/9/11/114005

[14] O'Mara FP. The significance of livestock as a contributor to global greenhouse gas emissions today and in the near future. Animal Feed Science and Technology. 2011;**166–167**:7–15. doi:10.1016/j.anifeedsci.2011.04.074

[15] Du Toit CJL, Meissner HH, Van Niekerk WA. Direct methane and nitrous oxide emissions of South African dairy and beef cattle. South African Journal of Animal Science. 2013;**43**(3):320–339. doi:10.4314/sajas.v43i3.7

[16] Hernandez-De Lira IO, Huber DH, Espinosa-Solares T, Balagurusamy N. Methane emission and bioenergy potential from livestock manures in Mexico. Journal of Renewable and Sustainable Energy. 2015;**7**(5):05117. doi:10.1063/1.4934564

[17] Castelán-Ortega OA, Ku-Vera JC, Estrada-Flores JG. Modeling methane emissions and methane inventories for cattle production systems in Mexico. Atmósfera. 2014;**27**(2):185–191. doi:10.1016/S0187-6236(14)71109-9

[18] Rendón-Huerta JA, Pinos-Rodríguez JM, García-López JC, Yáñez-Estrada LG, Kebreab E. Trends in greenhouse gas emissions from dairy cattle in Mexico between 1970 and 2010. Animal Production Science. 2014;**54**(3):292–298. doi:10.1071/AN12327

[19] Herrero M, Henderson B, Hvlik P, Thornton PK, Conant RT, Smith P, Wirsenius S, Hristov AN, Gerber P, Gill M, Butterbach-Bahl K. Greenhouse gas mitigation potentials in the livestock sector. Nature Climate Change. 2016;**6**:452–461. doi:10.1038/nclimate2925

[20] EPA. Environmental Protection Agency. Global Anthropogenic Non-CO2 Greenhouse Gas Emissions: 1990–2020. 2006. EPA Report No. 430-R-06-003, Office of Atmospheric Programs, Climate Change Division, Washington, DC.

[21] Russell JB, Rychlik JL. Factors that alter rumen microbial ecology. Science. 2001;**292**(5519):1119–1122. doi:10.1128/AEM.01289-09

[22] Mackie RI. Mutualistic fermentative digestion in the gastrointestinal tract: Diversity and evolution. Integrative and Comparative Biology. 2002;**42**(2):319–326. doi:10.1093/icb/42.2.319

[23] Valente TN, da Silva Lima E, dos Santos WB, Cesario AE, Tavares CA, de Freitas MA. Ruminal microorganism consideration and protein used in the metabolism of the ruminants: A review. African Journal of Microbiology Research. 2016;**10**(14):456–464. doi:10.5897/AJMR2016.7627

[24] Bergman EN. Energy contributions of volatile fatty acids from the gastrointestinal tract in various species. Physiological Reviews. 1990;**70**(2):567–590.

[25] Storm E, Ørskov ER. The nutritive value of rumen micro-organisms in ruminants. British Journal of Nutrition. 1983;**50**(2):479–485. doi:10.1079/BJN19830114

[26] Mackie RI, McSweeney CS, Aminov RI. Rumen. eLS. 2013. doi:10.1002/9780470015902.a0000404.pub2

[27] Patra AK. Enteric methane mitigation technologies for ruminant livestock: A synthesis of current research and future directions. Environmental Monitoring and Assessment. 2012;**184**(4):1929–1952. doi:10.1007/s10661-011-2090-y

[28] Gruninger RJ, Puniya AK, Callaghan TM, Edwards JE, Youssef N, Dagar SS, Filegerova K, Griffith GW, Forster R, Tsang A, McAllister T, Elshahed MS. Anaerobic fungi (phylum Neocallimastigomycota): Advances in understanding their taxonomy, life cycle, ecology, role and biotechnological potential. FEMS Microbiology Ecology. 2014;**90**(1):1–17. doi:10.1111/1574-6941.12383

[29] Kamra DN. Rumen microbial ecosystem. Current Science. 2005;**89**(1):124–135.

[30] Pers-Kamczyc E, Zmora P, Cieslak A, Szumacher-Strabel M. Development of nucleic acid based techniques and possibilities of their application to rumen microbial ecology research. Journal of Animal and Feed Sciences. 2011;**20**(3):315–337.

[31] Hungate RE. Ruminal Fermentation. In: Handbook of Physiology. Washington, DC: American Physiological Society. 1968, p. 2725–2745.

[32] Russell JB. Ecology of Rumen Microorganisms: Energy Use. In: Dobson A, Dobson MJ (eds.), Aspects of Digestive Physiology in Ruminants. Ithaca, NY: Cornell University Press. 1988, p. 74–98.

[33] Dijkstra J, Tamminga S. Simulation of the effects of diet on the contribution of rumen protozoa to degradation of fibre in the rumen. British Journal of Nutrition. 1995;**74**(5):617–634. doi:10.3168/jds.2006-518

[34] Dijkstra J, Boer H, Van Bruchem J, Bruining M, Tamminga S. Absorption of volatile fatty acids from the rumen of lactating dairy cows as influenced by volatile fatty acid concentration, pH and rumen liquid volume. British Journal of Nutrition. 1993;**69**(02):385–396. doi:10.1079/BJN19930041

[35] Morgavi DP, Forano E, Martin C, Newbold CJ. Microbial ecosystem and methanogenesis in ruminants. Animal. 2010;**4**(07):1024–1036. doi:10.1017/S1751731110000546

[36] Sieber JR, McInerney MJ, Gunsalus RP. Genomic insights into syntrophy: The paradigm for anaerobic metabolic cooperation. Annual Review of Microbiology. 2012;**66**:429–452. doi:10.1146/annurev-micro-090110-102844

[37] Abdoun K, Friederike S, Holger M. Ammonia and urea transport across the rumen epithelium: A review. Animal Health Research Reviews. 2006;**7**(1–2):43–59. doi:10.1017/S1466252307001156

[38] Hooke SE, Wright ADG, McBride BW. Methanogens: Methane producers of the rumen and mitigation strategies. Archaea. 2010;**2010**(1):1–11. doi:10.1155/2010/945785

[39] Johnson KA, Johnson DE. Methane emissions from cattle. Journal of Animal Science. 1995;**73**(8):2483–2492. doi:10.2527/1995.7382483x

[40] Zhou MI, Hernandez-Sanabria E. Assessment of the microbial ecology of ruminal methanogens in cattle with different feed efficiencies. Applied and Environmental Microbiology. 2009;**75**(20):6524–6533. doi:10.1128/AEM.02815-08

[41] Mao S, Zhang M, Liu J, Zhu W. Characterizing the bacterial microbiota across the gastrointestinal tracts of dairy cattle: Membership and potential function. Scientific Reports. 2015;**5**(16116):1–14. doi:10.1038/srep16116

[42] Henderson G, Cox F, Ganesh S, Jonker A, Young W. Rumen microbial community composition varies with diet and host, but a core microbiome is found across a wide geographical range. Scientific Reports. 2015;**5**:1–13. doi:10.1038/srep14567

[43] Pitta DW, Indugu N, Kumar S, Vecchiarelli B, Sinha R, Baker LD, Bhukya B, Ferguson JD. Metagenomic assessment of the functional potential of the rumen microbiome in Holstein dairy cows. Anaerobe. 2016;**38**:50–60. doi:10.1016/j.anaerobe.2015.12.003

[44] Kim M, Morrison M, Yu Z. Status of the phylogenetic diversity census of ruminal microbiome. FEMS Microbiology Ecology. 2011;**76**(1):49–63. doi:10.1111/j.1574-6941.2010.01029.x

[45] Wallace RJ, Rooke JA, McKain N, Duthie CA, Hyslop JJ, Ross DW, Waterhouse MW, Roehe R. The rumen microbial metagenome associated with high methane production in cattle. BMC Genomics. 2015;**16**(1):1–14. doi:10.1186/s12864-015-2032-0

[46] O'Neill BF, Deighton MH, O'loughlin BM, Mulligan FJ, Boland TM, O'donovan M, Lewis E. Effects of a perennial ryegrass diet or total mixed ration diet offered to spring-calving Holstein-Friesian dairy cows on methane emissions, dry matter intake, and milk production. Journal of Dairy Science. 2011;**94**(4):1941–1951. doi:10.3168/jds.2010-3361

[47] Li RW, Connor EE, Li C, Baldwin VI, Ransom L, Sparks ME. Characterization of the rumen microbiota of pre-ruminant calves using metagenomic tools. Environmental Microbiology. 2012;**14**(1):129–139. doi:10.1111/j.1462-2920.2011.02543.x

[48] Petri RM, Schwaiger T, Penner GB, Beauchemin KA, Forster RJ, McKinnon JJ, McAllister TA. Characterization of the core rumen microbiome in cattle during transition from forage to concentrate as well as during and after an acidotic challenge. PLoS One. 2013;**8**(12):e83424. doi:10.1371/journal.pone.0083424

[49] Beauchemin KA, McAllister TA, McGinn SM. Dietary mitigation of enteric methane from cattle. CAB Reviews: Perspectives in Agriculture, Veterinary Science, Nutrition and Natural Resources. 2009;**4**(9):1–18. doi:10.1079/PAVSNNR20094035

[50] Eckard RJ, Grainger C, De Klein CA. Options for the abatement of methane and nitrous oxide from ruminant production: A review. Livestock Science. 2010;**130**(1–3):47–56. doi:10.1016/j.livsci.2010.02.010

[51] Gerber PJ, Hristov AN, Henderson B, Makkar H, Oh J, Lee C, Meinen R, Montes F, Ott T, Firkins J, Rotz A. Technical options for the mitigation of direct methane and nitrous oxide emissions from livestock: A review. Animal. 2013;7(2s):220–234. doi:10.1017/S1751731113000876

[52] Hristov AN, Oh J, Firkins JL, Dijkstra J, Kebreab E, Waghorn G, Gerber PJ. Special topics—mitigation of methane and nitrous oxide emissions from animal operations: I. A review of enteric methane mitigation options. Journal of Animal Science. 2013;**91**(11):5045–5069. doi:10.2527/jas.2013-6583

[53] Kalac P. The effects of silage feeding on some sensory and health attributes of cow's milk: A review. Food Chemistry. 2011;**125**(2):307–317. doi:10.1016/j.foodchem.2010.08.077

[54] Beauchemin KA, Kreuzer M, O'mara F, McAllister TA. Nutritional management for enteric methane abatement: A review. Animal Production Science. 2008;**48**(8):21–27. doi:10.1071/EA07199

[55] Iqbal MF, Cheng YF, Zhu WY, Zeshan B. Mitigation of ruminant methane production: Current strategies, constraints and future options. World Journal of Microbiology and Biotechnology. 2008;**24**(12):2747–2755. doi:10.1007/s11274-008-9819-y

[56] Hegarty RS. Reducing rumen methane emissions through elimination of rumen protozoa. Crop and Pasture Science. 1999;**50**(8):1321–1328. doi:10.1071/AR99008

[57] Lettat A, Hassanat F, Benchaar C. Corn silage in dairy cow diets to reduce ruminal methanogenesis: Effects on the rumen metabolically active microbial communities. Journal of Dairy Science. 2013;**96**(8):5237–5248. doi:10.3168/jds.2012-6481

[58] Nagaraja TG, Titgemeyer EC. Ruminal acidosis in beef cattle: The current microbiological and nutritional outlook. Journal of Dairy Science. 2007;**90**:E17–EE38. doi:10.3168/jds.2006-478

[59] Bekele AZ, Koike S, Kobayashi Y. Genetic diversity and diet specificity of ruminal Prevotella revealed by 16S rRNA gene-based analysis. FEMS Microbiology Letters. 2010;**305**(1):49–57. doi:10.1111/j.1574-6968.2010.01911.x

[60] Saleem F, Ametaj BN, Bouatra S, Mandal R, Zebeli Q, Dunn SM, Wishart DS. A metabolomics approach to uncover the effects of grain diets on rumen health in dairy cows. Journal of Dairy Science. 2012;**95**(11):6606–6623. doi:10.3168/jds.2012-5403

[61] Loor JJ, Elolimy AA, McCann JC. Dietary impacts on rumen microbiota in beef and dairy production. Animal Frontiers. 2016;**6**(3):22–29. doi:10.2527/af.2016-0030

[62] Kittelmann S, Seedorf H, Walters WA, Clemente JC, Knight R, Gordon JI, Janssen PH. Simultaneous amplicon sequencing to explore co-occurrence patterns of bacterial, archaeal and eukaryotic microorganisms in rumen microbial communities. PLoS One. 2013;**8**(2):e47879. doi:10.1371/journal.pone.0047879

[63] Dumitru R, Palencia H, Schroeder SD, Demontigny BA, Takacs JM, Rasche ME, Ragsdale SW. Targeting methanopterin biosynthesis to inhibit methanogenesis. Applied and Environmental Microbiology. 2003;**69**(12):7236–7241. doi:10.1128/AEM.69.12.7236–7241.2003

[64] Zhou Z, Meng Q, Yu Z. Effects of methanogenic inhibitors on methane production and abundances of methanogens and cellulolytic bacteria in in vitro ruminal cultures. Applied and Environmental Microbiology. 2011;**77**(8):2634–2637. doi:10.1128/AEM.02779-10

[65] Denman SE, Tomkins NW, McSweeney CS. Quantitation and diversity analysis of ruminal methanogenic populations in response to the antimethanogenic compound bromochloromethane. FEMS Microbiology Ecology. 2007;**62**(3):313–322. doi:10.1111/j.1574-6941.2007.00394.x

[66] Tomkins NW, Colegate SM, Hunter RA. A bromochloromethane formulation reduces enteric methanogenesis in cattle fed grain-based diets. Animal Production Science. 2009;**49**(12):1053–1058. doi:10.1071/EA08223

[67] Mitsumori M, Shinkai T, Takenaka A, Enishi O, Higuchi K, Kobayashi Y, Nonaka I, Asanuma N, Denman SE, McSweeney CS. Responses in digestion, rumen fermentation and microbial populations to inhibition of methane formation by a halogenated methane analogue. British Journal of Nutrition. 2012;**108**(03):482–491. doi:10.1017/S0007114511005794

[68] Denman SE, Fernandez GM, Shinkai T, Mitsumori M, McSweeney CS. Metagenomic analysis of the rumen microbial community following inhibition of methane formation by a halogenated methane analog. Frontiers in Microbiology. 2015;**6**:1–12. doi:10.3389/fmicb.2015.01087

[69] Alvarado A, Montañez-Hernández LE, Palacio-Molina SL, Oropeza-Navarro R, Luévanos-Escareño MP, Balagurusamy N. Microbial trophic interactions and mcrA gene expression in monitoring of anaerobic digesters. Frontiers in Microbiology. 2014;**5**:597. doi:10.3389/fmicb.2014.00597

[70] Webster TM, Smith AL, Reddy RR, Pinto AJ, Hayes KF, Raskin L. Anaerobic microbial community response to methanogenic inhibitors 2-bromoethanesulfonate and propynoic acid. MicrobiologyOpen. 2016;**4**:537–550. doi:10.1002/mbo3.349

[71] Henderson G, Cook GM, Ronimus RS. Enzyme-and gene-based approaches for developing methanogen-specific compounds to control ruminant methane emissions: A review. Animal Production Science. 2016.A–J. doi:10.1071/AN15757

[72] Martinez-Fernandez G, Denman SE, Yang C, Cheung J, Mitsumori M, McSweeney CS. Methane inhibition alters the microbial community, hydrogen flow and fermentation response in the rumen of cattle. Frontiers in Microbiology. 2016;**7**:1–14. (1122). doi:10.3389/fmicb.2016.01122

[73] Hristov AN, Oh J, Giallongo F, Frederick TW, Harper MT, Weeks HL, Kindermann M. An inhibitor persistently decreased enteric methane emission from dairy cows with no negative effect on milk production. Proceedings of the National Academy of Sciences. 2015;**112**(34):10663–10668. doi:10.1073/pnas.1504124112

[74] Duin EC, Wagner T, Shima S, Prakash D, Cronin B, Yáñez-Rui DR, Kindermann M. Mode of action uncovered for the specific reduction of methane emissions from ruminants by the small molecule 3-nitrooxypropanol. Proceedings of the National Academy of Sciences. 2016;**113**(22):6172–6177. doi:10.1073/pnas.1600298113

[75] Romero-Perez A, Okine EK, McGinn SM, Guan LL, Oba M, Duval SM, Beauchemin KA. Sustained reduction in methane production from long-term addition of 3-nitrooxypropanol to a beef cattle diet. Journal of Animal Science. 2015;**93**(4):1790–1791. doi:10.2527/jas.2014-8726

[76] Bhatta R, Uyeno Y, Tajima K, Takenaka A, Yabumoto Y, Nonaka I, Enishi O, Kurihara M. Difference in the nature of tannins on in vitro ruminal methane and volatile fatty acid production and on methanogenic archaea and protozoal populations. Journal of Dairy Science. 2009;**92**(11):5512–5522. doi:10.3168/jds.2008-1441

[77] Reed JD. Nutritional toxicology of tannins and related polyphenols in forage legumes. Journal of Animal Science. 1995;**73**(5):1516–1528. doi:10.2527/1995.7351516x

[78] Makkar HPS, Francis G, Becker K. Bioactivity of phytochemicals in some lesser-known plants and their effects and potential applications in livestock and aquaculture production systems. Animal. 2007;**1**(09):1371–1391. doi:10.1017/S1751731107000298

[79] Grainger C, Clarke T, Auldist MJ, Beauchemin KA, McGinn SM, Waghorn GC, Eckard RJ. Potential use of Acacia mearnsii condensed tannins to reduce methane emissions and nitrogen excretion from grazing dairy cows. Canadian Journal of Animal Science. 2009;**89**(2):241–251. doi:10.4141/CJAS08110

[80] Jayanegara A, Goel G, Makkar HP, Becker K. Divergence between purified hydrolysable and condensed tannin effects on methane emission, rumen fermentation and microbial population in vitro. Animal Feed Science and Technology. 2015;**209**:60–68. doi:10.1016/j.anifeedsci.2015.08.002

[81] Castro-Montoya JM, Makkar HPS, Becker K. Chemical composition of rumen microbial fraction and fermentation parameters as affected by tannins and saponins using an in vitro rumen fermentation system. Canadian Journal of Animal Science. 2011;**91**(3):433–448. doi:10.4141/cjas2010-028

[82] Ghasemi S, Naserian AA, Valizadeh R, Tahmasebi AM, Vakili AR, Behgar M, Ghovvati S. Inclusion of pistachio hulls as a replacement for alfalfa hay in the diet of sheep causes a shift in the rumen cellulolytic bacterial population. Small Ruminant Research. 2012;**104**(1–3):94–98. doi:10.1016/j.smallrumres.2011.09.052

[83] Vincken JP, Heng L, de Groot A, Gruppen H. Saponins, classification and occurrence in the plant kingdom. Phytochemistry. 2007;**68**(3):275–297. doi:10.1016/j.phytochem.2006.10.008

[84] Wallace RJ, McEwan NR, McIntosh FM, Teferedegne B, Newbold CJ. Natural products as manipulators of rumen fermentation. Asian Australasian Journal of Animal Sciences. 2002;**15**(10):1458–1468. doi:10.5713/ajas.2002.1458

[85] Lila ZA, Mohammed N, Kanda S, Kamada T, Itabashi H. Effect of sarsponin on ruminal fermentation with particular reference to methane production in vitro. Journal of Dairy Science. 2003;**86**(10):3330–3336. doi:10.3168/jds.S0022-0302(03)73935-6

[86] Hess HD, Kreuzer M, Diaz TE, Lascano CE, Carulla JE, Soliva CR, Machmüller A. Saponin rich tropical fruits affect fermentation and methanogenesis in faunated and defaunated rumen fluid. Animal Feed Science and Technology. 2003;**109**(1):79–94. doi:10.1016/S0377-8401(03)00212-8

[87] Eugène M, Massé D, Chiquette J, Benchaar C. Meta-analysis on the effects of lipid supplementation on methane production in lactating dairy cows. Canadian Journal of Animal Science. 2008;**88**(2):331–337. doi:10.4141/CJAS07112

[88] Jenkins T. Success of fat in dairy rations depends on the amount. Feedstuffs (USA). 1997;**2:**11–12.

[89] Patra AK. The effect of dietary fats on methane emissions, and its other effects on digestibility, rumen fermentation and lactation performance in cattle: A meta-analysis. Livestock Science. 2013;**155**(2–3):244–254. doi:10.1016/j.livsci.2013.05.023

[90] Patra AK, Yu Z. Effects of essential oils on methane production and fermentation by, and abundance and diversity of, rumen microbial populations. Applied and Environmental Microbiology. 2012;**78**(12):4271–4280. doi:10.1128/AEM.00309-12

[91] Beauchemin KA, McGinn SM, Petit HV. Methane abatement strategies for cattle: Lipid supplementation of diets. Canadian Journal of Animal Science. 2007;**87**(3):431–440. doi:10.4141/CJAS07011

[92] Jordan E, Lovett DK, Hawkins M, Callan JJ, O'Mara FP. The effect of varying levels of coconut oil on intake, digestibility and methane output from continental cross beef heifers. Animal Science. 2006;**82**(06):859–865. doi:10.1017/ASC2006107.

Chapter 5

Public Health Aspect of Manure Management in Urban and Peri-Urban Livestock Farming in Developing Countries

Athumani Msalale Lupindu

Additional information is available at the end of the chapter

http://dx.doi.org/10.5772/65346

Abstract

Urban and peri-urban livestock farming has been expanding in recent decades due to high demand for animal proteins to feed the growing urban population. The increase in number of livestock and livestock keepers has led to increased manure production in a shrinking space. This chapter evaluates the risks of transmission of manure-borne pathogen between cattle, humans and the environment in urban and peri-urban areas. Cattle and manure management practices, government directives, the presence of zoonotic pathogens and risk of bacteria transmission were assessed by observations, interviews, bacteria isolation and characterization and statistical modeling. Cattle are kept under intensive and extensive systems. Different techniques are used to collect, convey, store and dispose manure, all of which lead to direct contact with humans. The prevalence of diarrheagenic *Escherichia coli* in cattle and water was 2.2% (95% CI: 0.99–3.67) and 0.5% (95% CI: 0.025–2.44), respectively. There was transmission of bacteria between cattle, humans and the environment in 52% of clusters. Cattle and manure management practices expose humans, livestock and the environment to risk of infection or contamination. Holistic approach can be adopted in this scenario to attain one health status and improve urban and peri-urban livestock contribution to community livelihood simultaneously.

Keywords: manure management, peri-urban, pathogen transmission, system thinking, one health

1. Introduction

Urban areas are city areas characterized by a dense human population of mixed age, sex, family and household structure, ethnic, cultural, religious diversity, educational and income levels, and high built-up area with technological and economic advancement. Rural areas, on the other hand, are open broad areas of land located far from towns and cities, which are composed of extensive bushes between large crop fields and livestock herds and sparse housing and population density. Between the urban and rural settings lies peri-urban zone whose population, livestock, crops and land use features are influenced by the proximate interface. Part of peri-urban area adjacent to urban area has features resembling urban features, while its other side assumes the rural characteristics. There is no distinct line separating the peri-urban from urban and rural settings, but a slow zone of change [1]. The gradual transition from peri-urban to urban setup is moving constantly away from city center toward the rural direction due to persistent urbanization pressure, especially in developing countries.

Urban and peri-urban livestock farming is expanding in developing countries primarily due to high demand for protein of animal origin to feed the rapidly growing urban populations, but also to generate income of livestock keeping households [2]. It is also a diversification tactic to spread livelihood risks in adverse situations [3]. Some urban and peri-urban dwellers continue to keep livestock to maintain their rural cultural values [4]. The expansion of urban and peri-urban livestock farming, which is reflected as an increase in number of both the livestock and households involved in keeping livestock, and rapid urban human population growth has increased the chance of contact between humans, animals and manure. Urban areas of Morogoro in Tanzania, for example, had a cattle population of 2618 in 1996 [5], which almost doubly increased to 4170 in 2006 [6]. By 2008, the cattle population in Morogoro urban was 19,099, and among them, 4425 were dairy cattle [7]. This cattle population hiked up to 49,625 in 2012 [8]. Rapid urban population growth is primarily caused by influx of people from rural areas either as migrants or as commuters [1, 9]. For instance, Tanzania's annual population growth rate between 1988 and 2002 was 3% with the urban population size increasing from 18% in 1988 to 23% in 2002 [10]. Moreover, population size and growth in rural and urban areas of Morogoro region from 2002 to 2012 show that the rural population grew by 23.7% from 1,279,513 in 2002 to 1,582,434 compared to 34.2% growth in urban population from 473,849 in 2002 to 636,058 in 2012 [8]. In this region, the general population density changed from 24 persons per square kilometer in 2002 to 31 persons per square kilometer in 2012 [11]. As a result of increased human and animal density, the chance of contact between humans and livestock has increased. The growth in animal population and concomitant increase in manure production, in shrinking space separating humans, livestock and manure, require appropriate livestock and manure management practices taking into account that livestock harbors zoonotic pathogens [12].

Four decades ago, before the expansion of urban and peri-urban livestock farming, free open communal cattle grazing system required minimal effort to manage manure [3, 13]. Cattle freely grazed during daytime and were confined during the night for security. Most manure was left scattered everywhere except for a small amount which was applied on crop fields [3, 14]. To date, the manure management practices have changed to adapt to densely populated areas where the space separating humans from animals and their wastes has decreased. A question arises: does this change consider prevention of animal and human from pathogen exposure as well as environmental contamination? This chapter describes assessment of manure management practices and risks of contact and transmission of cattle manure-related zoonotic pathogens between cattle, humans and the environment in urban and peri-urban areas of Morogoro region of Tanzania. This report forms a basis for developing strategies to improve urban and peri-urban livestock farming practices in order to safeguard human, animal and ecosystem health in settings similar to study area.

2. Exploration of animals-humans-environment interaction

Most of the people who keep cattle in urban and peri-urban areas also keep other livestock such as goats and chicken. Members of livestock keeping households share premises with livestock. In this community, livestock keeping households are randomly mixed with households which do not keep livestock. There is no tangible demarcation between livestock keeping households and non-livestock keeping households, and hence, the two types of households are in close contact. A total of 119 households keeping cattle, randomly selected, were willingly enrolled for the study in urban and peri-urban areas of Morogoro, Tanzania. Each cattle keeping household was paired to a non-cattle keeping household selected from any direction within a radius of 100 m for purpose of comparison. This pair was regarded as a cluster. Assessment of the interaction between cattle, humans and the environment within and between clusters involved field visits in order to make observations and interview household representatives about livestock and manure management practices. Questionnaire to cattle keeping households inquired about herd characteristics and management, manure management practices, awareness on zoonotic health risks and constraints to livestock farming in urban and peri-urban areas. Observations were made to top up and confirm the information gathered from the questionnaire. Details of labor division, herd composition and size, animal housing and feeding, herd health management, means and frequency of manure collection, storage and disposal were obtained at household level. Questionnaires to non-cattle keeping households enquired about attributes which may contribute to contact between humans, cattle and manure. Moreover, District Livestock Officers were interviewed about monitoring of manure handling practices in their respective areas of jurisdiction and were asked to present documents guiding livestock and manure management. This cross-sectional study was carried out from February 2010 to February 2012.

Cattle feces, human stool, soil and water samples were collected from each participating household for isolation and characterization of bacteria to check for the presence of pathogens and evidence of transmission between cattle, humans and the environment. In this particular study, *Escherichia coli* and non-typhoidal Salmonella spp. were target bacteria. Individual 100–150 g cattle fecal samples were collected by a gloved hand. A 100 g pooled soil sample from each household (cattle keeping and non-cattle keeping households) was obtained by taking 2–5 cm of top soil from five different areas within household premise. From each participating household, 100 ml water sample was collected in 250 ml container from stored water or sources such as boreholes, ponds or river which are used by humans and livestock. Stool sample from one household member was requested. For cattle keeping households, a member involved in cattle and/or manure management was eligible to give stool sample, while for non-cattle keeping households, any member was eligible. On the evening before sample collection day, a stool collection container was given to an appropriate person for collection of stool in the following morning. All samples for a cluster were collected on the same day and immediately placed in an insulated box with cooling elements and transported to the laboratory where bacteriological analysis was initiated.

Ethical clearance was approved by Sokoine University of Agriculture Ethical Committee to handle animals and animal samples. Approval was also obtained from the Tanzania National Institute for Medical Research (NIMR) Ethical Board (NIMR/HQ/R8a/Vol. IX/927) to handle and process human sample. All conditions for research approval were observed throughout the study.

3. Cattle and manure management practices

Observations and face-to-face interviews conducted during field visits to 119 households keeping cattle generated data which were analyzed by descriptive statistics such as means, frequencies and cross-tabulations by using SPSS 15.0. Information about herd characteristics and management, manure management practices and awareness on cattle manure-related zoonotic pathogens was obtained from the cattle keepers and Livestock Officers.

From observations and interviews, a total of 806 cattle were kept by study participants (minimum = 1, maximum = 36, mean = 7, median = 5, SD = 5.85), 95.8% of whom also kept animal species other than cattle in same residential premises. These animal species, with percentage of participants keeping these species in brackets, include chicken (80.7%), dogs (62.2%), goats (50.4%), pigs (27.7%), ducks (23.5%), cats (21.9%), sheep (10.9%), guinea fowls (9.2%), turkeys (5.9%), guinea pigs (1.7%), rabbits (1.7%) and monkey (0.8%). Cattle and manure management practices were carried out either by family members (46.2%) or by hired laborers (53.8%). Most cattle houses (71.4%) had concrete floor and the rest (28.6%) had floor made of earth. It was observed that majority of cattle houses (84%) had roofs and 16% were open cattle "boma." Cattle kept in earth floor houses with open or broken roof stayed on mud during rainy season. Three out of 119 respondents (2.5%) put grass on the floor of cattle house as bedding material, one of them had a house with earthed floor. All respondents kept their cattle in a confinement near to their residence for security reasons. Cattle were fed by "cut and carry" method under intensive system (47%) or were allowed to go around foraging (53%) where they mixed with livestock from other herds. There was sharing of water sources between cattle and humans. Free range cattle (40.3%) used surface water such as rivers, ponds and wells, while intensively kept cattle (59.7%) were provided water from taps also serving the people [15].

Overnight confinement of cattle resulted into manure accumulation which necessitated collection and storage/discard. Various methods were employed to collect, convey and store or discard manure. These included uses of utensils like spade, bucket or plastic bags, use of water splash and use of bare hands. Manure was collected by bare hands by a few respondents where there was direct contact with the manure. However, the majority of respondents used

http://dx.doi.org/10.5772/65346

utensils such as spades, hand hoes and rakes to collect manure into a pile within the cattle house. Some respondents used a water hose to collect manure (**Table 1**). Manure was removed from cattle house at different rates per day, week or month to storage or disposal site by using utensils (plastic bags, buckets, raw hides, spades and hand hoe and wheelbarrow), bare hands or water. The use of rubber boots was an observed practice by less than a half of the respondents, while the remaining fraction wore ordinary shoes, e.g., sandals or were barefooted while handling manure (**Table 1**). In all these different manure collection or conveyance methods, people did not use any protective measures such as special clothes or gloves and were observed to have direct skin contact with manure. A large proportion of respondents stored manure into piles before disposal as fertilizer or waste, whereas a few respondents threw fresh manure from cattle house direct into the surroundings. Most cattle keepers disposed manure within a radius of 10 m from their residential houses, especially those with land area of more than 1000 m^2. Respondents who did not spread manure on land opted for burning or giving it away to friends in plastic bags. Allowing effluent from cattle house to leach into immediate land was a common practice among cattle keepers although a few cattle keepers directed the effluent into a pit (**Table 1**).

Variable	Category	Frequency (%)
Manure disposal method	Spread on land	108 (90.8)
	Not spread on land	11 (9.2)
Means of manure collection	Hand picking	5 (4.2)
	Use of utensils	112 (94.1)
	Water splash	2 (1.7)
Frequency of manure collection	Once a day	72 (60.5)
	More than once a day	19 (16.0)
	Weekly	28 (23.5)
Means of manure conveyance	Hand picking	3 (2.6)
	Use of utensils	115 (96.6)
	Water splash	1 (0.8)
Use of rubber boots	Yes	70 (58.8)
	No	49 (41.2)
Manure treatment	Heaping	99 (83.2)
	Direct spread on land	20 (16.8)
Manure disposal distance	Within 10 m from residence	83 (69.7)
	Outside 10 m from residence	36 (30.3)
Effluent treatment	Direct spread on land	95 (79.8)
	Use of pit	24 (20.2)
Household area	>1000 m^2	87 (73.1)
	≤1000 m^2	32 (26.9)

Table 1. Manure management practices among 119 Morogoro urban and peri-urban cattle keepers.

Out of 119 respondents, 5% reported to have heard about manure-associated pathogens which can infect human. There were 125 responses to problems related to manure management which respondents encounter. Out of these 125 responses, 77 (61.6%) said they encounter no problem, while 15 (12%) responses reported that poor infrastructure impedes manure management practices. Lack of working facilities such as utensils and transport was reported in 13 (10.4%) responses as one of the problems cattle keepers face, whereas land scarcity appeared in 6 (4.8%) responses. Health problems related to respiratory tract, injuries and foot rot to manure handlers were mentioned in 5 (4%) responses, same as for the presence of poor cattle housing facilities. Odor and water scarcity were each mentioned in 2 (1.6%) responses as among problems of manure management practices in urban and peri-urban areas of Morogoro.

During the interview, Livestock Officers presented documents such as "Environmental Sanitation By-Laws" and "Animals in Urban areas By-Laws" which give directives on animal keep in the area and how to deal with wastes including manure. From interviews and the documents, the guideline which allows maximum herd size of four cattle per herd in urban area does not give area requirement specification and is not observed, and cattle manure is regarded by the by-laws and treated like any solid household waste [15]. It was observed that cattle keeping households are randomly distributed among non-cattle keeping households and there are no preconditions for a household to start keeping cattle. Anybody can start a herd of cattle anywhere in urban and peri-urban areas of Morogoro at any time.

The current manure management practices differ from those methods used a few decades ago in both the actual practices and resource base available which is shared between humans, animals and manure. Increased manure production in populated urban and peri-urban areas has resulted into the problems mentioned by cattle keepers. Some of these problems such as land scarcity odor and increased flies population have been previously reported to be due to exclusion of livestock farming during urban and peri-urban land use planning [4]. Increased manure production in a shrinking space has forced cattle keepers to collect, convey, store and finally dispose manure. Diverse cattle and manure management practices are determined by customs, convenience and availability of resources including land and equipment. Some farmers said that they keep cattle and handled the manure by the same methods since childhood; others opted for a particular cattle and manure management practice because it was easy to execute. Generally, there was direct contact between humans, cattle and manure and there was environmental contamination by fresh manure. In this scenario, humans, animals and environment are exposed to manure-related pathogens.

4. Pathogens in cattle, humans and the environment

Sample size of 100 clusters was calculated as previously described [16]. Face-to-face interviews were conducted to cattle keeping household members about cattle and manure management practices. Interview was also conducted to cattle keeping household neighbors who do not keep cattle about possible contact with cattle and manure. Individual fecal samples from 446 cattle, 100 stool samples from individuals keep cattle and 100 who do not keep cattle, 200 soil and 200 water samples from sources within homesteads were collected for bacteria isolation.

Escherichia coli was isolated and characterized as described earlier [16]. In summary, non-sorbitol fermenting (NSF) *E. coli* were isolated by using sorbitol MacConkey agar, and suspect colonies were characterized biochemically by use of MacConkey agar, Brilliance *E. coli* agar and indole test. Confirmed NSF *E. coli* isolates were assessed for the presence of virulence genes: intimin gene (*eae*), verocytotoxin 1 (*vtx1*), verocytotoxin 2 (*vtx2*), heat-stable enterotoxin, human variant (*estA*-human), heat-stable enterotoxin, porcine variant (*estA*-porcine), heat-labile enterotoxin (*eltA*) and invasive plasmid antigen (*ipaH*) by multiplex diarrheagenic *E. coli* (DEC) PCR. Dot-blot DNA hybridization was done by using *vtx1, vtx2, eae, ehxA, EAF, bfpA, saa, astA and vtx2f* DNA probes to confirm the presence of virulence genes in isolates positive by DEC PCR. The colonies were lysed, denatured and neutralized using standard conditions and then hybridized as formerly described [17].

Somatic antigen O and flagella antigen H on diarrheagenic *E. coli* were typed by using specific antisera at Statens Serum Institut, Denmark, using a standard protocol [18]. In summary, both somatic O and flagella H antigens were tested by agglutination method against both pooled and specific antisera. For somatic O antigen, a boiled culture of *E. coli* isolate was tested against a pooled O antisera and culture with positive agglutination test was further tested against single specific O antisera. Somatic O antigen was assigned a number according to positive agglutination on a specific single O antigen. For flagella H antigen, an *E. coli* culture was tested for motility in semi-solid medium and fixed with formaldehyde 0.5%. Fixed culture was tested against pooled H antisera, and positive culture was further tested against single specific H antisera. Fluffy reaction indicated positive result, and the isolates were assigned a number.

Phenotypic activity of virulence genes was assessed on Vero cell monolayers to test for cytopathic effects using protocol formerly described [17].

For non-typhoidal Salmonella spp. isolation, 1 ml of the sample suspension was enriched by overnight incubation in selenite fecal broth at 37°C. The bacteria growth was subcultured on Salmonella-Shigella agar at 37°C for 24 h. Colorless colonies with a black center were biochemically tested by urease and lysine carboxylase tests. Urease-negative and lysine carboxylase-positive colonies were tested against Salmonella polyvalent agglutinating sera (REMEL30858201 ZC02—LOT 820883) and serotyped by Kauffmann-White M03-03-001 method at Danish Institute for Technology (DTU).

Vero cytotoxin-producing *E. coli* (VTEC) from cattle, enteropathogenic *E. coli* (EPEC) from cattle and water and attaching and effacing *E. coli* (A/EEC) from cattle were isolated (**Figure 1**). Overall prevalence of diarrheagenic *E. coli* in cattle (n = 446) was 2.2% (95% CI 0.99–3.67) and in water (n = 200) was 0.5% (95% CI 0.025–2.44). The prevalence of VTEC in cattle was 1.6% (95% CI 0.69–3.08), (**Table 2**) [16].

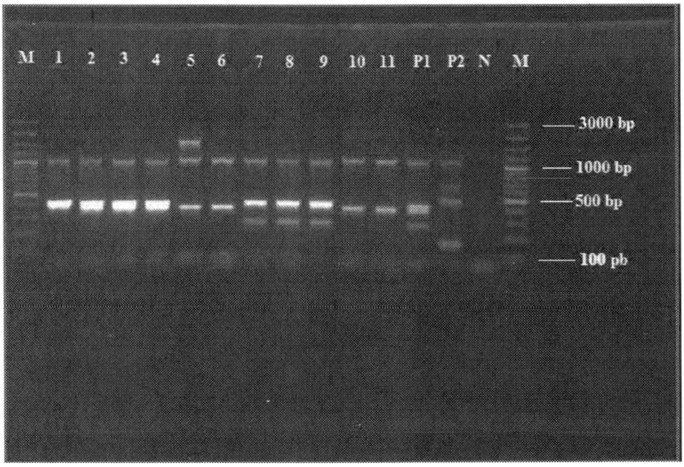

Figure 1. Multiplex DEC PCR for NSF *E. coli* isolates: lanes M: molecular weight size marker (100-bp plus DNA ladder); lane 1: *vtx2* and *eae*; lane 2: *vtx2* and *eae*; lane 3: *vtx2* and *eae*; lane 4: *vtx2* and *eae*; lane 5: *eae*; lane 6: *eae*; lane 7: *vtx1* and *vtx2*; lane 8: *vtx1* and *vtx2*; lane 9: *vtx1* and *vtx2*; lane 10: *eae*; lane 11: *eae*; lane P1: positive control for *vtx2*, *eae* and *vtx1*; lane P2: positive control for *ipaH, eltA* and *estA*; lane N: negative control.

Bacteria species	Source	Serotype	Pathotype	Virulence genes
Escherichia coli	Cattle	O157:H7	VTEC	vtx2, eae, ehxA and astA
	Cattle	O157:H7	VTEC	vtx2, eae, ehxA and astA
	Cattle	O157:H7	VTEC	vtx2, eae, ehxA and astA
	Cattle	O157:H7	VTEC	vtx2, eae, ehxA and astA
	Cattle	O113:H2	VTEC	vtx2
	Cattle	O+:H16	VTEC	vtx1 and vtx2
	Cattle	O113:H21	VTEC	vtx1 and vtx2
	Cattle	O142:H34	EPEC	eae, EAF and bfpA
	Water	O142:H34	EPEC	eae, EAF and bfpA
	Cattle	O+:H-	A/EEC	eae ehxA and astA
Salmonella kentucky	Cattle			
Salmonella kentucky	Cattle			
Salmonella weltevreden	Human			
Salmonella amager	Human			

Table 2. Zoonotic bacteria isolated from cattle, humans and environment in urban and peri-urban areas of Morogoro, Tanzania.

The prevalence of *Salmonella kentucky* in cattle was 0.45% (95% CI 0.001–0.016), while one *Salmonella weltevreden* and one *Salmonella amager* were isolated from different apparent healthy humans (**Table 2**).

The VTEC strains contained *vtx1a, vtx2b, vtx2c* and *vtx2d* subtypes either singly or in combinations, and phenotypic expression of virulence was confirmed by the cytopathic effect they caused to Vero cell monolayers (**Table 3**) [16].

Sample ID	Serotype	Source	VCA	*vtx1*	*vtx2*	*vtx*subtypes
BKIH101	0+:H16	Bovine	+	+	+	*vtx1a; vtx2c*
BKIN069	O157:H7	Bovine	–	–	+	*vtx2c*
BMKB070	O157:H7	Bovine	+	–	+	*vtx2c*
BMKB068	O157:H7	Bovine	+	–	+	*vtx2c*
BMKB069	O157:H7	Bovine	+	–	+	*vtx2c*
BMZU001	O113:H21	Bovine	+	–	+	*vtx2b + vtx2d*
BBIG020(1)	O113:H21	Bovine	+	+	+	*vtx1a; vtx2b + vtx2d*

Table 3. Vero cell assay (VCA) and *vtx* subtyping for non-sorbitol fermenting diarrheagenic *E. coli* isolates.

Isolation of diarrheagenic *E. coli* and Salmonella species from cattle feces is an evidence of risk of infection to humans and environmental contamination. There was also isolation of diarrheagenic *E. coli* from water in the study area. The risk in this scenario is due to direct contact between cattle, humans and manure as well as direct spread of fresh manure onto land within residence. This risk can cross between cattle keeping households because different cattle herds come into contact during grazing, and the spread can reach the non-cattle keeping neighbors. Sharing of water sources between humans and cattle, at some instances during dry season, poses another threat to public health. It is fortunate that these highly pathogenic and fatal diarrheagenic *E. coli* were not detected in humans because only apparent healthy subjects were sampled. Isolation of *Salmonella amager* and *S. weltevreden* in human stool calls for an attention on pathogen transmission route because humans can also act as a source of pathogens to livestock and the environment.

5. Transmission of bacteria between cattle, humans and environment

The study on transmission of bacteria involved 100 clusters, and each cluster was formed by a pair of a cattle keeping household and a neighboring non-cattle keeping household. Each cluster contributed two stool samples, two water samples and two soil samples, one of the samples from cattle keeping household and another from a non-cattle keeping household. Isolation, characterization and quantification of the risk of transfer of *E. coli* were done as earlier reported [19]. In summary, isolation of *E. coli* was carried out by inoculating a loopful suspension of cattle feces and stool from cattle keepers and non-cattle keepers, soil and water on

MacConkey agar followed by 24-h incubation at 37°C. *E. coli* suspected isolates were confirmed and screened for double antimicrobial resistance to ampicillin and tetracycline on antimicrobial embedded Petrifilm™ *Select E. coli* count (SEC) plate. Preparation of antimicrobial stock solution and screening procedure was done according to Ref. [20]. Ampicillin-tetracycline-resistant *E. coli* isolates were genetically assessed by pulsed-field gel electrophoresis (PFGE) according to Ref. [21]. Analysis and comparison of PFGE gel pictures were done by using GelCompar II software (Applied Maths, St-Martens-Latem, Belgium) as previously reported [18]. Isolates from cattle, humans, soil and water with 100% band pattern homology were considered genetically identical. A face-to-face interview was conducted to each household in the cluster. Semi-structured questionnaire which aimed at gathering information related to cattle and manure management (for cattle keeping households) and events or scenario leading to contact with cattle and manure (for non-cattle keeping households) was administered.

Logistic regression was run to quantify risk factors for the presence of isolates from cattle, humans, water or soil which are genetically identical to at least one other isolate from same or different clusters by using PROCGENMOD in SAS as earlier described [19]. The response variable was the occurrence of identical PFGE band pattern of *E. coli* isolates (yes or no), while the independent variables comprised of factors focusing on cattle herd characteristics and management (the presence of species other than cattle and labor division), cattle housing infrastructure (roof, floor and beddings), feeding and water system and manure management issues (collection and disposal). Univariable analysis was performed to all explanatory variables and those with an arbitrary p-value of equal or less to 0.25 were included in a multivariable model. A final model was obtained by a backward stepwise strategy. Chi-square test was used to check for association between different cattle and manure management factors at 5% significance level.

From 1046 samples, 118 (11.28%) samples produced ampicillin-tetracycline-resistant *E. coli*. Forty samples with resistant *E. coli* isolates (34%) were human stool, 50 (42%) were cattle feces, 21 (18%) were soil and 7 (6%) were water. One ampicillin-tetracycline-resistant *E. coli* isolate per sample was taken for further analyses. The 118 ampicillin-tetracycline-resistant *E. coli* isolates came from 44 out of the total 100 clusters. Twenty-three out of 44 clusters showing ampicillin-tetracycline-resistant isolates (52.3%) yielded at least one isolate with identical PFGE band pattern to another isolate from another source, suggesting that transfer of *E. coli* was a common event. Eight distinct PFGE band patterns designated arbitrary letters A, B, D, E, F, G, H and I for distinguishing purposes were identified. Inclusion of *Salmonella enterica* serovar *Braenderup* in all the gels showed a band pattern reproducibility of 100% (type C) (**Figure 2**) [19]. These PFGE band patterns cut across different clusters and were from cattle, humans, soil and water. Sixteen clusters out of 44 (36%) yielded at least one *E. coli* isolate which was identical to another isolate from another source by 100%. Seven clusters (16%) had isolate with similarity between 95 and 99.1% (**Figure 2**). PFGE band pattern A was comprised of five clusters, pattern B had two clusters, pattern D had three clusters, pattern E had six clusters, pattern F had two clusters, pattern G had one cluster, pattern H had two clusters and pattern I had also two clusters. Twelve isolates from cattle, human and soil constituted PFGE band pattern A, while pattern E was made up of eight isolates from cattle, soil and water (**Table 4**).

http://dx.doi.org/10.5772/65346

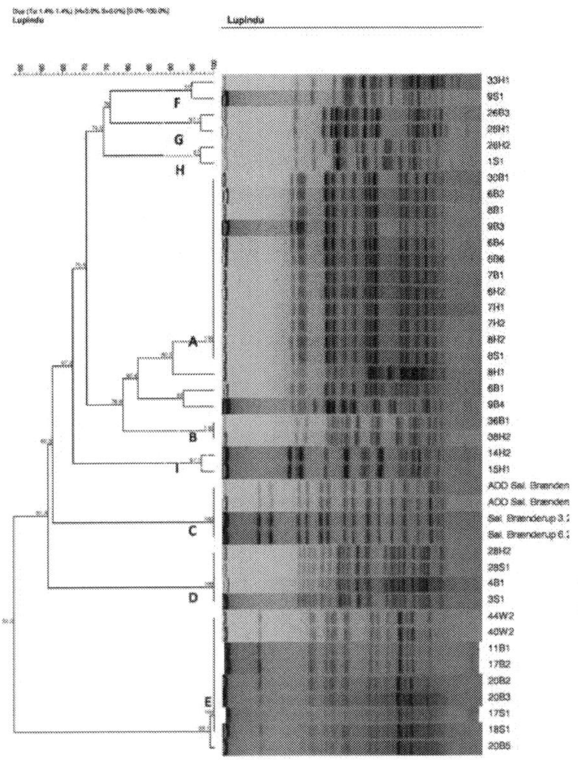

Figure 2. PFGE band pattern for ampicillin- and tetracycline-resistant *E. coli* isolates from humans, cattle, soil and water.

This shows that there was sharing of genetic characteristics between bacteria isolates from different sources. There was also genetic relatedness in cluster seven between isolates from cattle keeping human (7H1), cattle (7B2) and non-cattle keeping human (7H2). This scenario suggests that sharing of bacteria go beyond cattle keeping households to their non-cattle keeping neighbors. In some instance, like in cluster six, isolates from cattle (6B2, 6B4 and 6B6) did not resemble humans in the same household, but had PFGE band pattern identical to neighboring non-cattle keeping human (6H2). Sharing of genetic features was also observed in isolates from cattle, humans and the environment. For instance, isolate from cattle in cluster eight (8B1) was identical to isolate from non-cattle keeping human (8H2) and isolate from soil collected from cattle keeping household (8S1) in the same cluster eight. In PFGE band pattern E, isolates from water sources of non-cattle keeping households (40W2 and 44W2) had identical PFGE patterns to isolates from cattle (11B1, 17B2, 20B2 and 20B3) and soil (17S1 and 18S1) from cattle keeping households (**Table 4**). Some isolates with identical PFGE band patterns from cattle, e.g., in PFGE band pattern A, came from different households/herds, signifying the role of communal grazing in sharing of bacteria between cattle.

Isolates with distinct PFGE band patterns within clusters had a good temporal relationship in terms of sampling and isolation. Most of them came from samples collected on one day or within a week (**Table 4**) [19].

Clonal group	Cluster	Isolate ID[a]				Sample date
A	6	6B2	6B4	6B6	6H2	16 July 2011
	7	7B2	7H1	7H2		16 July 2011
	8	8B1	8H2	8S1		20 July 2011
	9	9B3				20 July 2011
	30	30B1				20 July 2011
B	36	36B1				21 September 2011
	38	38H2				21 September 2011
D	28	28H2	28S1			15 September 2011
	4	4B1				15 September 2011
	3	3S1				22 July 2011
E	11	11B1				24 September 2011
	17	17B2	17S1			15 January 2012
	18	18S1				15 January 2012
	40	40W2				15 January 2012
	20	20B2	20B3			18 January 2012
	44	44W2				11 January 2012
F	33	33H1				20 July 2011
	9	9S1				20 July 2011

[a]*E. coli* isolates from humans (H), water (W) and soil (S) with odd last digit originated from cattle keeping households while those with even last digit were obtained from non-cattle keeping neighbors.

Table 4. Identical PFGE patterns of ampicillin- and tetracycline-resistant *E. coli* isolated from cattle keeping and non-cattle keeping neighbor households in peri-urban areas of Morogoro, Tanzania.

Escherichia coli isolates from cattle were found in all clusters with identical PFGE bands patterns (**Figure 2**), proposing that cattle are the focal point of bacteria sharing and manure is the center of contact between cattle, humans and the environment. These roles of cattle and manure in bacteria sharing between cattle, humans and the environment lead to a hypothetical bacteria transmission pathways presented in **Figure 3**. The bacteria sharing pathways can be used to set up strategies to break the contact and transmission pathways. However, there is a need to develop procedures which can be used to determine the donor-recipient bacteria transmission relationship, something that was not done in the current study.

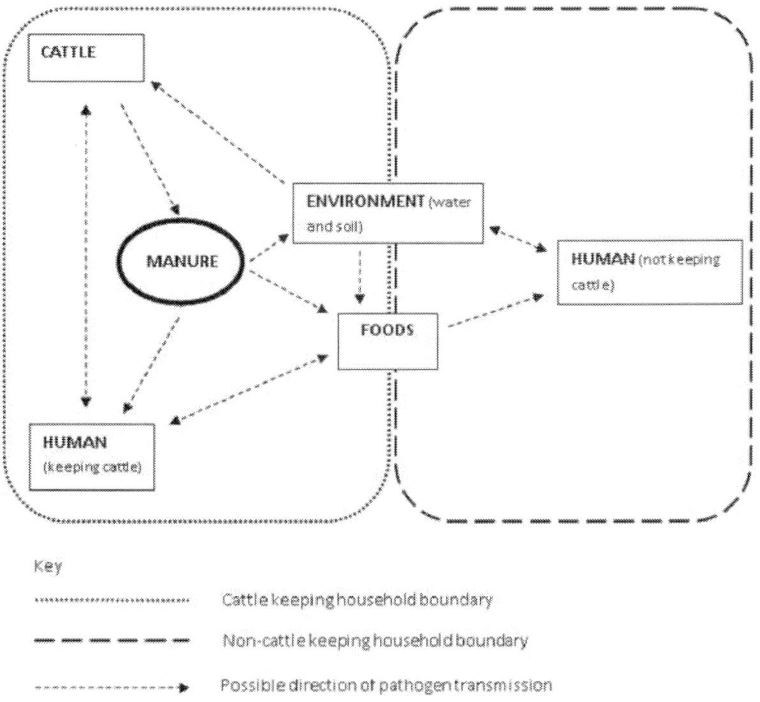

Figure 3. Hypothetical transmission pathways of enteric bacteria in urban and peri-urban livestock farming systems in Morogoro, Tanzania.

From univariable analysis, five explanatory variables, namely manure responsible personnel (family member or hired laborer), cattle house roof (present or absent), cattle house floor (concrete or earth), use of bedding (yes or no) and animal water source (tap or surface water) qualified and progressed to multivariable logistic regression analysis. There were no detected confounders during the model building process, and the final logistic regression model was made up of a single explanatory variable, the type of cattle house roof. The cattle house with a roof was at 11 times odds of having isolates with identical PFGE band pattern to another isolate from another source (OR = 11.2, 95% CI 1.1–119.3). Generally, isolates with PFGE band pattern identical to at least one isolate from another source were 33, 86.8% of which were isolated from cattle houses with a roof. The model goodness-of-fit test, expressed as the ratio of deviance to degree of freedom, was 1.2, while the correlation of 0.1344 existed between sample sources from different clusters. This shows that the variables were well explained by the model.

From this study, it seems that there was transmission of bacteria in roofed cattle houses than in cattle houses without roof. This could be due to the effect of direct sun rays in open cattle houses killing the bacteria before the transmission.

Cattle feeding system was statistically associated with cattle water sources ($X^2 = 28.5$, df = 1, $p \leq 0.0001$), whereby free range cattle used surface water and cattle under zero grazing used tap water which was also used by humans. On the other hand, distance from residence to manure disposal site was statistically associated with the way manure was handled ($X^2 = 8$, df = 1, $p = 0.005$). That is, cattle keeping households which stored manure in heaps disposed manure within residential areas, whereas households which opted to spread fresh manure on land did it outside residential area [19].

6. Conclusion

Cattle and manure management practices in urban and peri-urban livestock farming allow direct contact of cattle manure with humans, cattle and the environment. Humans and cattle are at risk of infection with enteric pathogens and the environment to contamination because enteric pathogens have been isolated from fresh cattle feces in urban and peri-urban areas. Under the current manure management system, there is transmission of commensal enteric bacteria between cattle, humans and the environment (water and soil), in which case, same route can transmit enteric pathogens. The risk of human and livestock infection and environment contamination is potentiated by the fact that cattle keepers are unaware of such manure-related pathogens and majority of them do not perceive that there are public health threats from the current cattle and manure management practices. The risk of enteric pathogen transmission to humans extends beyond cattle keeping households to their non-cattle keeping neighbors. Current cattle and manure management practices in urban and peri-urban areas of Morogoro put the whole community (cattle keepers and non-cattle keepers), cattle and other domestic animals, at risk of infection and the environment (water and soil) to contamination.

7. Recommendations

The reported public health challenges can be alleviated by adopting a system thinking or holistic approach, whereby all stakeholders are identified and involved, at their respective capacities, in planning, execution and monitoring of urban and peri-urban livestock farming. This approach will aim at safeguarding public, livestock and ecosystem health at the same time improving urban and peri-urban livestock contribution toward community livelihood. Some of the key stakeholders, each of whom may have a different key role in ensuring this goal is achieved, include personnel from health section, agriculture, livestock, local government authorities, land use planning, civil engineers, environmental conservation, demography, law enforcing sections, politicians and the general public. For example, local government authorities may put preconditions for starting a cattle herd in urban and peri-urban areas and set criteria for maintenance of livestock keeping permit. This procedure may facilitate other livestock-related activities such as disease control, surveillance and traceability of animals. Moreover, land use planning and environmental conservation sections may set specific areas for keeping livestock, while medical and veterinary sections may jointly control zoonoses. The law enforcing personnel can facilitate in making sure regulations related to livestock, and livestock products are observed. The general public should be well informed of and participate in control of manure-related zoonotic pathogens. The holistic setup of urban and peri-urban livestock farming should take into account all the features of continuous change from rural to peri-urban to urban setting. This means that planning of livestock farming in peri-urban should suit the urban setup even when the peri-urban area is urbanized.

To reduce human and animal contact with manure and to reduce the risk of human and animal infection and environmental contamination, the following strategies are recommended.

- Urban and peri-urban land use planning should include livestock industry during planning so that specific areas are legally recognized for livestock farming in urban and peri-urban areas.
- There should be strategies to convert manure into a convenient, safe and valuable commodity. This should involve reduction in water content and odor from manure while maintaining its soil fertilizing quality.
- Education to community (livestock keepers and non-livestock keepers) on livestock and manure-related zoonotic risks which are associated with management practices. It should be the responsibility of the whole community to ensure one health status is achieved.
- Appropriate regulations, by-laws and guidelines should be formulated and reinforced to guide safe cattle and manure management practices which safeguard public, livestock and ecosystem health. The guidelines should clearly give directives on personal protection, cattle and manure handling and environmental protection.

Acknowledgements

The author wishes to thank the Ministry of Foreign affairs of Denmark through the Danish International Development Agency (Danida) for financial support to the project "Opportunities and challenges in peri-urban livestock farming in Tanzania" (P6-08-Tan), to which this work is a part. The excellent collaboration between Sokoine University Agriculture (Tanzania) and University of Copenhagen (Denmark) during implementation of this study is highly appreciated. Special recognition is granted to Statens Serum Institut (Denmark) for their technical support during laboratory work. I acknowledge participation and project support of the community (cattle keepers and non-cattle keepers) and Livestock Extension Officers in urban and peri-urban areas of Morogoro.

Author details

Athumani Msalale Lupindu

Address all correspondence to: amsalale@yahoo.com; alupindu@suanet.ac.tz

Department of Veterinary Medicine and Public Health, College of Veterinary and Medical Sciences, Sokoine University of Agriculture, Morogoro, Tanzania

References

[1] Simon D., McGregor D., Nsiah-Gyabaah K. The changing urban-rural interface of African cities: Definitional issues and an application to Kumasi, Ghana. Environment and Urbanization. 2004;16(2):235–248. doi:10.1177/095624780401600214

[2] Mlozi M.R.S. Impacts of urban agriculture in Dar es Salaam, Tanzania. The Environmentalist. 1997;17(2):115–124.

[3] Guendel S., editors. Peri-urban and urban livestock keeping in East Africa—a coping strategy for the poor? Aylesford: Department of International Development (DFID); 2002. 32 p.

[4] Mvena Z.S.K. The past, present and future of urban agriculture in Tanzania. Journal of Agricultural Economics and Development. 1999;3:71–77.

[5] United Republic of Tanzania. Morogoro region socio-economic profile. Dar es Salaam: The Planning Commission and Regional Commissioner's Office Morogoro; 1997.

[6] United Republic of Tanzania. Morogoro regional socio-economic profile. Morogoro: Morogoro Regional Commissioner's Office; 2007.

[7] United Republic of Tanzania. National sample census of agriculture 2007/2008. Dar es salaam: National Bureau of Statistics; 2012. 331 p.

[8] United Republic of Tanzania. Basic demographic and socio-economic profile report Tanzania mainland. Dar es Salaam: National Bureau of statistics; 2014. 237 p.

[9] Briggs J. The Peri-urban zone of Dar es Salaam, Tanzania: recent trends and changes in agricultural land use. Transactions of the Institute of British Geographers. 1992;16(3): 319–331.

[10] United Republic of Tanzania. Population and housing census. Dar es Salaam: National Bureau of Statistics; 2012.

[11] United Republic of Tanzania. Tanzania in figures. Dar es Salaam: National Bureau of Statistics; 2013.

[12] Smith K.E., Stenzel S.A., Bender J.B., Wagstrom E., Soderlund D., Leano F.T. et al. Outbreak of enteric infections caused by multiple pathogens associated with calves at a farm day camp. The Pediatric Infectious Disease Journal. 2004;23(12):1098–1104. doi: 10.1097/01.inf.0000145409.74116.e5

[13] Pell A.N. Manure and microbes: public and animal health problem? Journal of Diary Science. 1997;80(10):2673–2681.

[14] Mlozi M.R.S. Urban agriculture: ethnicity, cattle raising and some environmental implications in the city of Dar es Salaam, Tanzania. African Studies Review. 1997;40(3): 1–28.

[15] Lupindu A.M., Ngowi H.A., Dallsgaard A., Olsen J.E., Msoffe P.L.M. Current manure management practices and hygiene aspects of urban and Peri-urban livestock farming in Tanzania. Livestock Research for Rural Development. 2012;24:Article #167.

[16] Lupindu A.M., Olsen J.S., Ngowi H.A., Msoffe P.L.M., Mtambo M.M., Scheutz F., Dalsgaard A. Occurrence and characterization of Shiga toxin-producing *Escherichia coli* O157:H7 and other non-sorbitol-fermenting *E. coli* in cattle and humans in urban areas of Morogoro, Tanzania. Vector-borne and Zoonotic Diseases. 2014;14(7):503–510. doi: 10.1089/vbz.2013.1502

[17] Scheutz F. Vero cytotoxin producing *Escherichia coli* (VTEC) isolated from Danish patients [thesis]. Copenhagen: Statens Serum Institut; 1997.

[18] Bergan T, editor. Methods in microbiology. 1st ed. London: Academic Press; 1984. 375 p.

[19] Lupindu A.M., Dalsgaard A., Msoffe P.L.M., Ngowi H.A., Mtambo M.M., Olsen J.E. Transmission of antibiotic-resistant *Escherichia coli* between cattle, humans and the environment in peri-urban livestock keeping communities in Morogoro, Tanzania. Preventive Veterinary Medicine. 2015;118(4):477–482. doi:10.1016/j.prevetmed. 2014.12.005

[20] Wu S., Chouliara E., Jensen L.B., Dalsgaard A. Evaluation of Petrifilm™ select *E. coli* count plate medium to discriminate antimicrobial resistant *Escherichia coli*. Acta Veterinaria Scandinavica. 2008;50(1):38. doi:10.1186/1751-0147-50-38

[21] Ribot E.M., Fair M.A., Gautom R., Cameron D.N., Hunter S.B., Swaminathan B. et al. Standardization of pulsed-field gel electrophoresis protocol for the subtyping of *Escherichia coli* O157:H7, Salmonella and Shigella for PulseNet. Foodborne Pathogen and Disease. 2006;3(1):57–67. doi:10.1089/fpd.2006.3.59

Chapter 6

Lactation Responses toward Milk Indigenous Enzymes

Aneela Hameed, Majid Hussain and Saeed Akhtar

Additional information is available at the end of the chapter

http://dx.doi.org/10.5772/66392

Abstract

Milk being a highly nutritious food in its natural form provides energy. There are various factors influencing the composition of milk: breed, stage of lactation, nutritional status, health, and milking intervals. A number of indigenous enzymes present in milk are being affected by stages of lactation period. Their concentration varies during early, mid and late lactation periods. This varied behavior ultimately affects the quality of dairy products. In this chapter, the level of milk enzymes: lipases and esterases, plasmin (PL), plasminogen (PLG) phosphatases (alkaline phosphatase ALP; acid phosphatase (ACP), lysozyme (LZ), lactoperoxidase (LP), xanthine oxidoreductase (XOR), and catalase (CAT) will be reviewed with respect to the stages of lactation periods.

Keywords: milk, indigenous enzymes, lactation stages, parity, season

1. Introduction

Milk is one of the perfect, complete, and primitive dairy food known by mankind. It is white and nutritious physiological secretion from the mammary glands of mammals, serves as nourishment for their neonates [1, 2]. It is a major product obtained from healthy and highly productive dairy animals. Physiological and nutritional provisions of each species are more or less distinctive. The breed, health, nutritional status, stage of lactation period, and milking intervals are some of the factors that affect the milk composition [3, 4]. The variation in constituents occurs entire lactation period. Lactation stage is the prime factors that affect the milk properties and some of the enzyme activities [5, 6]. Solids-non-fat (SNF) content is frequently highest throughout first 2–3 weeks of lactation.

2. Milk indigenous enzymes

Numerous enzymes have been indigenously identified in milk from 1924 to 1970 [7]. A large number of enzymes with multiple functionalities are present in milk. Additional enzymes contribute in quality of milk products and also perform an antibacterial action (LP). In bovine milk more than 70 enzymes are detected [8, 9]. A 50–60 substantial number of milk enzymes with multiple functions are present in abundance in milk and are concerned with processing stability and general customer safety [10] and additionally processing suitability (ALP). Some enzymes (LP) having antibacterial characteristics are with significant importance in preservation of milk and milk products and some, e.g., plasmin (PL) and lipoprotein lipase (LPL) connected with the serum, plasma, fat globules, casein, or leukocytes are important in maintaining of quality of milk and milk products. More than 40 enzymes have been recognized in cow milk [9, 11].

Lactation period in animals involves colostrum, developed milk, peak, and production with compositional variations. Numerous indigenous enzymes present in milk are secreted by epithelial cells and their composition changes with the lactation stages.

In already recognized indigenous enzymes in milk [7], almost 20 enzymes have been well characterized and the rest of the 40 enzymes are of little significance but can be identified through their activity. These enzymes indicate the efficient process of milk pasteurization (ALP, γ-glutamyl transferase GGT) or of mastitis (phosphatases, CAT). Additional enzymes can be of significance in processing and ultimately providing safety to human beings. They play an antibacterial activity (LP) and contribute quality to milk products (e.g., LPL, PL) associated with the serum, plasma, fat globule, casein, or leukocytes.

2.1. Lipases and esterases

Lipolytic enzymes have capability to hydrolyze triacylglycerols are considered as carboxylesterases [12, 13]. Those enzymes that can hydrolyze acyl glycerol having <10 carbon atom fatty acids are known as esterases or carboxylases (Enzyme Commission, EC 3.1.1.1) while those can hydrolyze ≥10 carbon atom fatty acids are considered as lipases, or triacylglycerol acyl hydrolases (EC 3.1.1.3) [14, 15].

Esterases are different from lipases due to their functions for being relatively soluble compared to emulsified ester substrates. Several esterases are present in milk [15, 16], the most prominent are carboxylesterase (EC 3.1.1.1), acetylcholinesterase (EC 3.1.1.7), and cholinesterase (EC 3.1.1.8). In bovine colostrum, lipase is not connected with casein and not activated by blood serum, therefore exhibited low lipase activity and showed slight lipolysis in early lactation. However, after few days of calving, normal milk from early lactation exhibited higher lipase activity [17, 18].

Lipases are naturally a critical group of enzymes since they are connected with the fat digestion system. Lipases are more dynamic at pH 8–9 and catalyze the advancement of hydrolytic rancidity in milk. Investigation of lipases is more alluring in the light of the fact that it would add to our comprehension about the properties and modes of these enzymes [19, 20].

The phenomenon of lipolysis is correlated with the lactation days. Higher activity is associated with its presence in fat fraction of milk. Activity of lipase in milk fat increases with the advancement in lactation stages [21]. The lipolysis process is of major apprehension in the dairy industry, as rancid off flavors are produced in milk and milk products during this phenomenon [22].

Earlier research has well established that milk lipase is sensitive to heavy metals. Copper, cobalt, and nickel have been shown to be more powerful inhibitors of lipase than iron, chromium, manganese, and silver. Enzyme activity is stimulated by blood serum albumin, ammonium, calcium ions, and mercaptoethanol. The buffer solutions, citrate, acetate, and phosphate buffers damage the enzyme activity, whereas borate and barbiturate buffers do not [23, 24].

LPL in cow milk is altered due to the breed, lactation phase, feed and fodder, season, and milk yield [22, 25]. Lipase activity increased from 0.32 to 2.98 U/mL of milk. At the point when milk fat globule membrane (MFGM) is damaged, lipolysis takes place rapidly and leads to hydrolytic rancidity and ultimately may cause variations in functionality and flavor of dairy products throughout storage period [15]. LPL found in goat milk is of low concentration in the early and late lactation stage [26].

The membrane lipase is available in higher concentration in milk from dairy animals in late lactation [27]. They additionally reported that lipase action in milk showed inclined pattern with reference to lactation stages. Hameed et al. [28] reported the expanding pattern of lipase activity with lactation stages in bovine milk. Lipase action (1.55 U/mL) was recorded higher ($p < 0.01$) in milk, examined at the last of lactation, followed by other lactations (1.29 and 1.16 U/mL, respectively).

2.2. Plasmin

Plasmin (PL; EC 3.4.21.7) is an alkaline serine proteinase enzyme that proteolytically cleaves the blood clots [29]. This enzyme has affinity toward arginine (Arg) and lysine (Lys) residues, specifically breaks the Arg-X and Lys-X bonds [30, 31]. The activity of enzyme is increased with the multiple factors that include lactation stage, lactation number and severity of mastitis infection [32–34].

On the basis of origin, the PL and plasminogen (PLG) are considered to be migrated from blood to milk, and higher activity of PL in peak lactation designating more conversion of PLG into PL in bovine milk [35–37].

PL is basically released in the form of PLG in normal milk. The concentration of PLG (0.8–2.8 μg/mL) in fresh milk is varied and its concentration is 2–30 times higher than that of PL (0.1–0.7 μg/mL). It is activated by storage or when milk is stayed in the lumen of mammary glands before milking. A considerable interest has been involved in the activation of PLG, upon which activity of PL depends [38–41]. It promptly hydrolyzes the bonding of β-casein, α_{s2}-casein, and α_{s1}-casein and affects the quality of dairy products [32, 42].

Advancing lactation stage is an essential factor that influences PL activity and percentage, suggesting that more PL activity in milk from goat and older cows is a result of increased PLG activation [43–45]. However, the relevant information about the varied concentration and activity of PL during lactation stages is controversial. Leitner et al. [46] declared significantly higher activities of PL in infected glands of sheep.

Caropresе et al. [47] and Albenzio et al. [33] found that there was decrease in PL activity in ewe's milk from the early to the late lactation stage whereas Koutsouli et al. [48] and Bianchi et al. [49] announced that PL activity significantly affected by udder health status and found an increased level of PL activity due to more somatic cell counts (SCCs) during the late lactation period.

The variation in PLG-derived activity and total PL plus PLG-derived activity is greatly influenced by lactation stage and seasonal changes. It is linked with reduction of milk yield and advancement in lactation stage [45, 50, 51]. Due to increased activity of plasminogen, more entry of PL occurred from blood to milk inside the mammary glands [52]. The PL and PLG activities were significantly increased in the advancement of lactation and a nonsignificant decrease in their ratio (PL:PLG) was observed as compared to camel milk [53, 54].

2.3. Phosphatases

2.3.1. Alkaline phosphatase

In 1925, for the first time, phosphatase enzyme in milk is documented by Demuth and then considered as an alkaline phosphatase (ALP; EC 3.1.3.1) indigenous to milk by Graham and Kay [55]. It became recognizable when it was confirmed that the requirement for time-temperature relationships to inactivate the ALP required slightly higher as compared to kill *Mycobacterium tuberculosis* [56, 57]. Almost 40% activity of ALP in raw cow milk is declared to be linked with the milk fat globule membrane (MFGM) in the cream phase, though the rest is soluble or dispersed in whey membrane particles (WMP) in skim milk [58]. Between individuals and herds, higher ALP levels vary significantly and its concerned activity is correlated with lactation stages and mastitis [59, 60]. Magnesium and zinc ions are promoter of ALP while tin, copper, cobalt, and ethylenediaminetetraacetic acid (EDTA) have inhibitory action and iron has no effect on activities of ALP [61].

ALP activity is in inverse relationship with yield but the other factors, e.g., fat content, breed, and feed, have no effect. For ovine milk, the ALP content is contrarily linked with milk production and directly to the milk fat substance, while infected milk (mastitis) has higher ALP activity [62, 63]. It is reported that ALP activity is low at the mature milk production stage, increased to maximum activity during the peak production stage and again decreased at the end production stage [28]. ALP activity in cow increases as lactation stage proceeds. Immediately after parturition, there is a decrease in ALP activity with a further sharp decrease after the first milking period. ALP activity then continues to decrease and noted minimum at the first week. Then increased slowly and found maximum by the 28th week of lactation [64]. In another study, ALP activity in milk was found lowest in the early lactation stages and progressed along with advancement of lactation stages and milk yields decreased. These ALP activities were also noticed greater in milk samples from evening milk as compared with morning milk [65].

2.3.2. *Acid phosphatase*

Acid phosphomonoesterase (ACP; EC 3.1.3.2) in milk was initially identified by Huggins and Talalay [66] and affirmed by Mullen [67], declared that ACP was ideally in the active form at 4.0 pH. It was thermally stable and for complete inactivation it required 88°C for 10 min. ACP in bovine milk hydrolyzes the phosphate group of casein particles [68]. There are some components that act as inhibitor and activator. Fluoride acts as an inhibitor for ACP activity but slightly activated by Mn^{2+}. In milk, the ACP level is just ~2% that of the ALP level. Approximately 75% of ACP was found generally in the skim milk phase and 20% of ACP in the MFGM [68, 69]. Reducing agents, ascorbic acid and 2-mercaptoethanol increases the ACP activity by 100% in skim milk, whereas the ACP activity in MFGM is unaffected by these agents. Casein acts as a substrate for the activity of ACP and major casein fractions αs ($\alpha s1 + \alpha s2$) > β > κ also serve as competitive inhibitors as the ACP enzyme binds with the phosphate group of casein. The ability to bind calcium with κ-casein to form micelles is reduced by dephosphorylation of casein [61].

ACP in milk might be of innovative significance due to three reasons. First, ACP exhibits thermal stability and because of this property it may be used as an indicator for severe heat treatment rather than normal. Second, numerous milk items may have a pH near to that of its optimum. Third, phosphoproteins such as caseins might be dephosphorylated readily. Technological milk properties and development of dairy products depend on the integrity of casein micelles. The enhanced activity of ACP may create problem in the inactivation of ACP without affecting nutritional qualities as it is linked with gelation of ultra-high temperature (UHT) and development of cheese flavor [70, 71].

Specific activity of ACP is greater in cream; however, about 80% ACP of milk is present in skimmed milk [60]. ACP levels in milk of Sahiwal dairy animals showed a declined pattern alongside lactation stages [28]. Shakeel-ur-Rehman and Farkye [72] observed the higher activity of ACP at 5–6 days postpartum, and afterward observed declined trend up to the end of lactation stage. Nevertheless, the range of ACP levels in their study was presented from 2.6 ± 10^{-4} to 2.6 ± 10^{-3} U/mL in normal cow milk. The ACP level is 4–10 times more in mastitis milk than normal cow milk [73, 74].

2.4. Lysozyme

Lysozyme (LZ; EC 3.2.1.17; muramidase) is a single polypeptide chain (14.3 KDa M.W.), cross-linked by four disulfide bonds [75, 76]. It is an important bacteriolytic protein in milk, component of the antibacterial system, that kills bacteria by cleaving the β-1,4-glycosidic bond between N-acetyl muramic acid and N-acetyl glucosamine residues in peptidoglycan of the bacterial cell wall [77, 78].

It helps in improving the human health status, especially neonate, to protect them from infections of invading pathogens with the promotion of gut microbiota until their own immune system is developed [79–81].

Basically, there are two types of LZ: hen egg-white (C-LZ) and goose egg-white (G-LZ). However, both C-LZ and G-LZ forms may be present in cow milk as these forms are present in other body fluids and in stomach tissue of the cow [82].

LZ is available at higher concentration (0.420 g/L) in human milk as compared to buffalos (3.85 μg/mL), cow (0.0013 g/L), and goat (0.0025 g/L) milk [83–86].

The activity of LZ was in greater extent and more stable in buffalo milk as compared to cow milk. However, colostrum possessed 5 times higher activity as compared to mature milk. It was also observed that various factors: parity of animal and lactation stage not influences the activity of LZ but it was increased during the peak summer and winter seasons [86–88]. A substantial increase of milk LZ in mastitis among different bovine species suggested that the neutrophils are the most probable source of LZ due to inflammation of mammary gland [89–91]

2.5. Lactoperoxidases

Lactoperoxidase (LP; EC 1.11.1.7) is the second most abundant enzyme after xanthine oxidase in bovine milk. The most generally prescribed industrial utilization of LP systems is the preservation of raw fresh milk during transportation and storage in dairy plants [92, 93]. It received a considerable attention as an optimum indicator of super-pasteurized milk [94]. Its level in bovine milk is about 30 mg/L constituting approximately 1% whey protein [95]. The LP system (LP-thiocyanate-H_2O_2) is a natural preservation system and has antimicrobial characteristics. Oxidation of thiocyanate in the presence of H_2O_2 is catalyzed by activated LP and produce hypothiocyanate (OSCN) or higher oxides (antimicrobial compounds). These compounds exhibited their antimicrobial properties by oxidizing the sulfhydryl groups of proteins to disulfides [96].

LP enzyme activities are affected by various factors, i.e., sexual cycle, season, lactation, diet, and breed [95, 97]. LP activity in bovine milk (1.2–19.4 U/mL) is about 20 times higher in peroxidase action than human milk [98]. LP levels in dairy animals milk is ranged from 1.5 to 2.7 U/mL with a general mean of 2.3 U/mL [99]. The LP level is low in colostrum of dairy animals, after that adopted inclined trend rapidly after 3–5 days postpartum [95]. LP enzyme activity is a precursor to diagnose the mastitis disease in dairy animals. The activity of LP increases as the somatic cell count (SCC) increases [100, 101].

The LP activity of cows adopted declined trend along with lactation stages. The activity of LP decreases with the advancement in lactation stages (9.64–6.66 U/mL) [28]. The decreasing trend along with lactation stages was also observed by Althaus et al. [102] who reported significant reduction in LP activity from the early stage of lactation toward the end of lactation. Reiter [95] observed a significant increase in LP activity between 4 and 5 days after calving of the lactation period, followed by a gradual decrease toward close of lactation. The reduction in action of LP activity in cow milk could be due to increase in the thiocyanate content as Fonteh et al. [99] described that the LP level promoted with an increase in 2,2'- azino-bis (3-ethylbenzothiazoline-6-sulphonic acid) or ABTS contents but reduced with an increase in thiocyanate contents. They also reported that LP activity is enhanced with whey protein, lactose, magnesium, sodium, and calcium chlorides, and reduced in occurrence of casein.

2.6. Catalases

Catalase (CAT; H_2O_2:H_2O_2 oxidoreductase; EC 1.11.1.6) dismutates hydrogen peroxide (H_2O_2) into water (H_2O) and free oxygen (O_2) [8, 103]. CAT was among the first enzymes present in milk. Babcock and Russell [104] portrayed that an extract of separator slime can break down H_2O_2. The CAT activity in milk fluctuates with feedstuff and lactation phase, level expanded particularly during mastitis [103, 105]. CAT has the ability to degrade the surplus hydrogen peroxide and reduce oxidative infection caused by reactive oxygen species (ROS) [106].

CAT and SCC contributed in the mastitis risk markers. Risk level of mastitis and losses in milk production increase with the advancement in parity, phase of lactation, and also in spring and winter seasons [107]. Measurement of CAT activity plays a distinct role in monitoring the health status of udders in cow. The antioxidant activity of enzyme CAT increases when SCC increases [100, 108, 109]. CAT antioxidant activity is higher in colostrum, then reductions occur as the lactation stage proceeds and again high in the late lactation [110, 111]. Its absence in milk is an indication of an efficient pasteurization process [7].

2.7. Xanthine oxidoreductase

Xanthine oxidoreductase (XOR; EC 1.13.22; 1.1.1.204) is a milk indigenous enzyme having capability of oxidizing hypoxanthine to xanthine and xanthine to uric acid with the reduction of O_2 to H_2O_2 [7, 112]. This protein is initially presented in milk; in 1902, Schardinger reported that this compound is competent for oxidizing aldehydes to acids by the lessening the methylene blue and after that generally called this chemical as "Schardinger enzyme." XOR has been established to require FAD^+ and Mo^{++} for its optimum catalytic action [103, 113, 114].

XOR is concentrated in MFGM, which is the second most abundant protein constituting, 20% of the MFGM protein. Milk is a good source of XOR, some of its portion is shifted to mammary glands by means of the blood circulation system. The XOR level in milk differs recognizably during lactation. However, bovine milk contains significant levels of XOR (1.4–1.8 U/mg) as compared to goat (0.27 U/mg) and sheep (0.69 U/mg) milk and camel (nd) milk because enzyme molecules lack molybdenum (Mo^{++}) [115–118]. This level can be amplified by complementing the diet with Mo^{++} [7].

In buffalo milk, XOR (0.75 U/mg) exists in the catalytically inactive form because of higher concentration of demolybdo and desulfo forms. Structural factors and lower contents of Fe/S might be the possible reason of lowering enzymatic activity of XOR in buffalo [119].

Surprisingly, camel milk exhibited no detectable XOR activity and its Mo^{++} contents were comparable to human and goat milk [120].

Being significant part of lactating cells, the levels of XOR mRNA began to increase during mid-pregnancy, turned upward at the onset of lactation and diminished quickly in constrained involution [121]. XOR expression remained constant, while specific activity enhanced at the initial lactation phase that facilitates in milk synthesis [122]. Physiologically, XOR contains hydrogen peroxide, nitric oxide, and superoxide ion, mainly functions as in the activation of various metabolic pathways [123]. XOR contributes to an antimicrobial defense mechanism in the gastrointestinal tract (GIT) tract and plays a significant role in the immune system of mammary glands [111, 124, 125]. XOR activity increases during infectious diseases and its cytotoxic action is useful for the defenses against bacteria [123].

3. Conclusions

Conclusively, intensive review of enzyme activities has revealed the significance of indigenous milk enzymes with varied concentration behavior during lactation stages. Lactation stage has a prominent effect on enzymes activities and ultimately it may affect the technological behavior of milk composition.

Generally, colostrum formation contains higher enzyme activities than during the established lactation period. Mastitis or several other progressions that increase leukocytes in milk increase enzyme activities such as CAT. LP, ACP, and LP decrease while lipase activity increases with progress of lactation. ALP activity first increases then decreases at the end of lactation. PL activity increases in the late lactation because of that it makes milk less suitable for cheese making.

This varied behavior of enzyme activities at early, mid, and late lactation stages can be a troubling problem for manufacturing of milk and milk products in various regions of the world. As enormous animals in late lactation periods and considerable seasonal variations affect the ultimate quality of milk and have a better increased choice to process the specified valued dairy product. Furthermore, milk from mid lactation would be a balanced source of energy to maintain the health status of the individuals.

Author details

Aneela Hameed, Majid Hussain* and Saeed Akhtar

*Address all correspondence to: mhussain@bzu.edu.pk

Institute of Food Science & Nutrition, Faculty of Agricultural Sciences, Bahauddin Zakariya University, Multan, Pakistan

References

[1] Pehrsson P, Haytowitz D, Holden J, Perry C, Beckler D. USDA's national food and nutrient analysis program: food sampling. J. Food Compos. Anal. 2000;13:379–389. doi: 10.1006/jfca.1999.0867

[2] Nickerson S. Milk production: Factors affecting milk composition. In: Harding F, editor. Milk Quality: Springer, USA; 1995. p. 3–24. doi: 10.1007/978-1-4615-2195-2_2

[3] Misra S, Sharma A, Bhattacharya T, Kumar P, Roy S. Association of breed and polymorphism of αs1- and αs2-casein genes with milk quality and daily milk and constituent yield traits of buffaloes (*Bubalus bubalis*). Buffalo Bulletin. 2008;27:294–301.

[4] Fox P, Guinee T, Cogan T, McSweeney P. Fundamentals of Cheese Science. Aspen Publishers, Inc., Gaithersburg, MD; 2000.

[5] Looper ML. Factors Affecting Milk Composition of Lactating Cows. Division of Agriculture, University of Arkansas System, USA, 2012.

[6] Schutz M, Hansen L, Steuernagel G, Kuck A. Variation of milk, fat, protein, and somatic cells for dairy cattle. J. Dairy Sci. 1990;73:484–493. doi: 10.3168/jds.S0022-0302(90)78696-1

[7] Fox P, Kelly A. Indigenous enzymes in milk: overview and historical aspects—Part 1. Int. Dairy J. 2006;16:500–516. doi: 10.1016/j.idairyj.2005.09.013

[8] Fox PF. Significance of indigenous enzymes in milk and dairy products. Food Science and Technology, Marcel Dekker, New York; 2003. pp.255–278.

[9] Kelly A, Fox P. Indigenous enzymes in milk: a synopsis of future research requirements. Int. Dairy J. 2006;16:707–715. doi: 10.1016/j.idairyj.2005.10.018

[10] Moatsou G. Indigenous enzymatic activities in ovine and caprine milks. Int. J. Dairy Technol. 2010;63:16–31. doi: 10.1111/j.1471-0307.2009.00552.x

[11] Walstra P, Jenness R. Dairy Chemistry & Physics. John Wiley & Sons, New York, USA; 1984.

[12] Zhao W-S, Hu S-L, Yu K, Wang H, Wang W, Loor J, Luo J. Lipoprotein lipase, tissue expression and effects on genes related to fatty acid synthesis in goat mammary epithelial cells. Int. J. Mol. Sci. 2014;15:22757–22771. doi: 10.3390/ijms151222757

[13] Beisson F, Tiss A, Rivière C, Verger R. Methods for lipase detection and assay: a critical review. Eur. J. Lipid Sci. Technol. 2000;102:133–153. doi: 10.1002/(SICI)1438-9312(200002)102:2<133::AID-EJLT133>3.0.CO;2-X

[14] Chahiniana H, Sarda L. Distinction between esterases and lipases: comparative biochemical properties of sequence-related carboxylesterases. Protein Pept. Lett. 2009;16:1149–1161. doi: 10.2174/092986609789071333

[15] Chen L, Daniel RM, Coolbear T. Detection and impact of protease and lipase activities in milk and milk powders. Int. Dairy J. 2003;13:255–275. doi: 10.1016/S0958-6946(02)00171-1

[16] Acton QA. Esterases—Advances in Research and Application: 2013 edition. Scholarly Editions, Atlanta, Gorgia; 2013.

[17] Fuquay JW, Fox PF, McSweeney PL. Encyclopedia of Dairy Sciences. Academic Press, San Diego; 2011. doi: 10.1016/B978-0-12-374407-4.09006-3

[18] Saito Z, Kim G. Effects of lactation stage on temperature-activated lipolysis and lipase activity in cow's milk. Japanese J. Dairy Food Sci. (Japan). 1994; 44:A139–145.

[19] Deeth H, Fitz-Gerald C. Lipolytic enzymes and hydrolytic rancidity. Advanced Dairy Chemistry Volume 2 Lipids. Springer, New York; 2006. pp. 481–556. doi: 10.1007/0-387-28813-9_15

[20] Palmquist D. Milk fat: origin of fatty acids and influence of nutritional factors thereon. Advanced Dairy Chemistry Volume 2 Lipids. Springer, New York; 2006. pp. 43–92. doi: 10.1007/0-387-28813-9_2

[21] Ahrné L, Björck L. Lipolysis and the distribution of lipase activity in bovine milk in relation to stage of lactation and time of milking. J. Dairy Res. 1985;52:55–64. doi: 10.1017/S002202990002389X

[22] Deeth HC. Lipoprotein lipase and lipolysis in milk. Int. Dairy J. 2006;16:555–562. doi: 10.1016/j.idairyj.2005.08.011

[23] Chandan R, Shahani K. Milk lipases. A review. J. Dairy Sci. 1964;47:471–480. doi: 10.3168/jds.S0022-0302(64)88694-X

[24] Egelrud T, Olivecrona T. The purification of a lipoprotein lipase from bovine skim milk. J. Biol. Chem. 1972;247:6212–6217.

[25] Toušová R, Stádník L, Ducháček J. Effect of season and the time of milking on spontaneous and induced lipolysis in bovine milk fat. Czech. J. Food Sci. 2013;31:20–26.

[26] Chilliard Y, Ferlay A, Rouel J, Lamberet G. A review of nutritional and physiological factors affecting goat milk lipid synthesis and lipolysis. J. Dairy Sci. 2003;86:1751–1770. doi: 10.3168/jds.S0022-0302(03)73761-8

[27] Tarassuk N, Frankel E. The specificity of milk lipase. IV. Partition of the lipase system in milk. J. Dairy Sci. 1957;40:418–430. doi: 10.3168/jds.S0022-0302(57)94496-X

[28] Hameed A, Hussain R, Zahoor T, Akhtar S, Riaz M, Ismail T. Effect of oxytocin on enzyme activities in bovine milk. Int. Dairy J. 2014;39:229–231. doi: 10.1016/j.idairyj.2014.06.013

[29] Turner RB, Liu L, Sazonova IY, Reed GL. Structural elements that govern the substrate specificity of the clot-dissolving enzyme plasmin. J. Biol. Chem. 2002;277:33068–33074. doi: 10.1074/jbc.M203782200

[30] Fox P, McSweeney P. Proteolysis in cheese during ripening. Food Rev. Int. 1996;12:457–509. doi: 10.1080/87559129609541091

[31] Kitchen BJ. Indigenous milk enzymes. In: Fox P, editor. Developments in Dairy Chemistry—3: Springer, Netherlands; 1985. p. 239–279. doi: 10.1007/978-94-009-4950-8_9

[32] Ismail B, Nielsen S. Invited review: plasmin protease in milk: current knowledge and relevance to dairy industry. J. Dairy Sci. 2010;93:4999–5009. doi: 10.3168/jds.2010-3122

[33] Albenzio M, Caroprese M, Santillo A, Marino R, Muscio A, Sevi A. Proteolytic patterns and plasmin activity in ewes' milk as affected by somatic cell count and stage of lactation. J. Dairy Res. 2005;72:86–92. doi: 10.1017/S0022029904000676

[34] Bastian ED, Brown RJ. Plasmin in milk and dairy products: an update. Int. Dairy J. 1996;6:435–457. doi: 10.1016/0958-6946(95)00021-6

[35] Di Luccia A, Picariello G, Trani A, Alviti G, Loizzo P, Faccia M, Addeo F. Occurrence of β-casein fragments in cold-stored and curdled river buffalo (Bubalus bubalis L.) milk. J. Dairy Sci. 2009;92:1319–1329. doi: 10.3168/jds.2008-1220

[36] Farrell H, Jimenez-Flores R, Bleck G, Brown E, Butler J, Creamer L, Hicks C, Hollar C, Ng-Kwai-Hang K, Swaisgood H. Nomenclature of the proteins of cows' milk—Sixth revision. J. Dairy Sci. 2004;87:1641–1674. doi: 10.3168/jds.S0022-0302(04)73319-6

[37] Kaminogawa S, Mizobuchi H, Yamauchi K. Comparison of bovine milk protease with plasmin. Agric. Biol. Chem. 1972;36:2163–2167. doi:/10.1271/bbb1961.36.2163

[38] Crudden A, Kelly AL. Studies of plasmin activity in whey. Int. Dairy J. 2003;13:987–993. doi: 10.1016/S0958-6946(03)00140-7

[39] Nielsen S. Plasmin system in milk. In: Roginski H, Fuquay J, Fox P, editors. Encyclopedia of dairy sciences. II: Bodmin, Cornwall, UK; 2003. p. 929–934. doi: 10.1016/B0-12-227235-8/00153-X

[40] Ozen BF, Hayes KD, Mauer LJ. Measurement of plasminogen concentration and differentiation of plasmin and plasminogen using Fourier-transform infrared spectroscopy. Int. Dairy J. 2003;13:441–446. doi: 10.1016/S0958-6946(03)00055-4

[41] Nielsen SS. Plasmin system and microbial proteases in milk: characteristics, roles, and relationship. J. Agric. Food Chem. 2002;50:6628–6634. doi: 10.1021/jf0201881

[42] Crudden A, Fox F, Kelly AL. Factors affecting the hydrolytic action of plasmin in milk. Int. Dairy J. 2005;15:305–313. doi: 10.1016/j.idairyj.2004.08.008

[43] Cortellino G, Locci F, Rampilli M. An investigation of the plasmin–plasminogen system in caprine milk and cheese. Int. Dairy J. 2006;16:619–622. doi: 10.1016/j.idairyj.2005.11.012

[44] Politis I. The role of plasminogen activator in the bovine mammary gland. In: Mol J, Clegg R, editors. Biology of the Mammary Gland. 480: Springer, USA; 2002. p. 203–207. doi: 10.1007/0-306-46832-8_25

[45] Bastian ED, Brown RJ, Ernstrom CA. Plasmin activity and milk coagulation. J. Dairy Sci. 1991;74:3677–3685. doi: 10.3168/jds.S0022-0302(91)78557-3

[46] Leitner G, Chaffer M, Shamay A, Shapiro F, Merin U, Ezra E, Saran A, Silanikove N. Changes in milk composition as affected by subclinical mastitis in sheep. J. Dairy Sci. 2004;87:46–52. doi: 10.3168/jds.S0022-0302(04)73140-9

[47] Caroprese M, Schena L, Marzano A, Marino R, Sevi A, Albenzio M. Contribution of macrophages to plasmin activity in ewe bulk milk. Ital. J. Anim. Sci. 2007;6:545–547. doi: 10.4081/ijas.2007.1s.545

[48] Koutsouli P, Patounas G, Massouras T, Bizelis I, Politis I. Plasmin–plasminogen system and milk coagulation properties of two Greek dairy sheep breeds. Small Ruminant Res. 2015;124:89–94. doi: 0.1016/j.smallrumres.2015.01.015

[49] Bianchi L, Bolla A, Budelli E, Caroli A, Casoli C, Pauselli M, Duranti E. Effect of udder health status and lactation phase on the characteristics of Sardinian ewe milk. J. Dairy Sci. 2004;87:2401–2408. doi: 10.3168/jds.S0022-0302(04)73362-7

[50] Nicholas GD, Auldist MJ, Molan PC, Stelwagen K, Prosser CG. Effects of stage of lactation and time of year on plasmin-derived proteolytic activity in bovine milk in New Zealand. J. Dairy Res. 2002;69:533–540. doi: 10.1017/S0022029902005745

[51] Politis I, Hang KNK, Giroux R. Environmental factors affecting plasmin activity in milk. J. Dairy Sci. 1989;72:1713–1718. doi: 10.3168/jds.S0022-0302(89)79286-9

[52] Richardson B, Pearce K. Determination of plasmin in dairy products. N.Z. J. Dairy Sci. Technol. 1981;16:209–220.

[53] Hamed H, Trujillo A-J, Juan B, Guamis B, ElFeki A, Gargouri A. Interrelationships between somatic cell counts, lactation stage and lactation number and their influence on plasmin activity and protein fraction distribution in dromedary (Camelus dromedaries) and cow milks. Small Ruminant Res. 2012;105:300–307. doi: 10.1016/j.smallrumres.2012.01.002

[54] Chaiyabutr N, Thammacharoen S, Komolvanich S, Chanpongsang S. Effects of long-term administration of recombinant bovine somatotropin on the plasminogen–plasmin system and milk composition of crossbred Holstein cattle. Anim. Sci. J. 2007;78:251–258. doi: 10.1111/j.1740-0929.2007.00432.x

[55] Graham W, Kay H. Phosphorus compounds of milk. V. The phosphorus partition in milk, with preliminary observations on milk phosphatase. J. Dairy Res. 1933;5:54–62. doi: 10.1017/S0022029900000923

[56] Rankin S, Christiansen A, Lee W, Banavara D, Lopez-Hernandez A. Invited review: the application of alkaline phosphatase assays for the validation of milk product pasteurization. J. Dairy Sci. 2010;93:5538–5551. doi: 10.3168/jds.2010-3400

[57] Rosenthal I, Bernstein S, Rosen B. Alkaline phosphatase activity in *Penicillium roqueforti* and in blue-veined cheeses. J. Dairy Sci. 1996;79:16–19. doi: 10.3168/jds.S0022-0302(96)76328-2

[58] Silanikove N, Shapiro F. Distribution of xanthine oxidase and xanthine dehydrogenase activity in bovine milk: physiological and technological implications. Int. Dairy J. 2007;17:1188–1194. doi: 10.1016/j.idairyj.2007.03.003

[59] Babaei H, Mansouri-Najand L, Molaei M, Kheradmand A, Sharifan M. Assessment of lactate dehydrogenase, alkaline phosphatase and aspartate aminotransferase activities in cow's milk as an indicator of subclinical mastitis. Vet. Res. Commun. 2007;31:419–425. doi: 10.1007/s11259-007-3539-x

[60] Andrews A. Indigenous enzymes in milk: General introduction. In: Fox P, editor. Advanced dairy chemistry-1: Proteins. II ed: Elsevier Science Publishers; 1992. p. 286–292.

[61] Fox PF, McSweeney PL. Dairy Chemistry and Biochemistry. Blackie Academic & Professional, an imprint of Chapman & Hall, London; 1998.

[62] Anifantakis E, Rosakis P. Alkaline phosphatase activity of sheep's milk and some factors affecting it. Egypt. J. Dairy Sci. 1983;11:173–182.

[63] Patil M, Nagvekar A, Ingole S, Bharucha S, Palve V. Somatic cell count and alkaline phosphatase activity in milk for evaluation of mastitis in buffalo. Vet. World. 2015;8:363. doi: 10.14202/vetworld.2015.363-366

[64] Haab W, Smith L. Variations in alkaline phosphatase activity of milk. J. Dairy Sci. 1956;39:1644–1650. doi: 10.3168/jds.S0022-0302(56)94905-0

[65] Banks J, Muir D. Variability of alkaline phosphatase in goat milk in relation to its use as an effective index of pasteurisation. Project S01003. Hannah Research Institute, Ayr, Scotland. 2004.

[66] Huggins C, Talalay P. Sodium phenolphthalein phosphate as a substrate for phosphatase tests. J. Biol. Chem. 1945;159:399–410.

[67] Mullen J. 427. The acid phosphatase of cows' milk: I. Some properties of the enzyme. J. Dairy Res. 1950;17:288–305. doi: 10.1017/S0022029900005847

[68] Akuzawa R, Fox PF. Acid phosphatase in cheese. Anim. Sci. J. 2004;75:385–391. doi: 10.1111/j.1740-0929.2004.00202.x

[69] Flynn N, Akuzawa R, Fox P, editors. Characterisation of isozymes of acid phosphatase isolated from bovine milk. Abstracts 28th Food Science and Technology Research Conference. University College, Cork; 1998. p. 165.

[70] Fox P, Uniacke-Lowe T, McSweeney P, O'Mahony J. Heat-induced changes in milk. Dairy Chemistry and Biochemistry: Springer International Publishing, Switzerland; 2015. p. 345–375. doi: 10.1007/978-3-319-14892-2_9

[71] Fox PF, Morrissey PA. Indigenous Enzymes of Bovine Milk. In: Birch GG, Blakebrough N, Parker KJ, editors. Enzymes and Food Processing. 3rd ed: Springer Netherlands; 2012. p. 213–238. doi: 10.1007/978-94-011-6740-6

[72] Shakeel-ur-Rehman, Farkye NY. Enzymes indigenous to milk: phosphatase. In: Roginski H, Fuquay J, Fox P, editors. Encyclopedia of Dairy Sciences: Amsterdam Academic Press; 2002. p. 934–938. doi: 10.1016/B0-12-227235-8/00154-1

[73] Fox P, Stepaniak L. Enzymes in cheese technology. Int. Dairy J. 1993;3:509–530. doi: 10.1016/0958-6946(93)90029-Y

[74] Andrews A. Further studies on acid phosphatase of leucocyte origin in normal and mastitic bovine milks. J. Dairy Res. 1976;43:127–131. doi: 10.1017/S0022029900015673

[75] Cegielska-Radziejewska R, Lesnierowski G, Kijowski J. Properties and application of egg white lysozyme and its modified preparations—a review. Pol. J. Food Nutr. Sci. 2008;58. doi: 10.1007/s00217-010-1347-y

[76] Masschalck B, Michiels CW. Antimicrobial properties of lysozyme in relation to foodborne vegetative bacteria. Crit. Rev. Microbiol. 2003;29:191–214. doi: 10.1080/713610448

[77] Wiesner J, Vilcinskas A. Antimicrobial peptides: the ancient arm of the human immune system. Virulence. 2010;1:440–464. doi: 10.4161/viru.1.5.12983

[78] Park YW, Haenlein GF. Handbook of Milk of Non-Bovine Mammals. John Wiley & Sons, New York; 2008. doi: 10.1002/9780470999738

[79] Al-Jabri AA. Honey, milk and antibiotics. Afr. J. Biotechnol. 2005;4:1580–1587.

[80] Lönnerdal B. Nutritional and physiologic significance of human milk proteins. Am. J. Clin. Nutr. 2003;77:1537S-1543S.

[81] Maga EA, Sargent RG, Zeng H, Pati S, Zarling DA, Oppenheim SM, Collette NM, Moyer AL, Conrad-Brink JS, Rowe JD. Increased efficiency of transgenic livestock production. Transgenic Res. 2003;12:485–496. doi: 10.1023/A:1024257906647

[82] Losnedahl K, Wang H, Aslam M. Antimicrobial proteins in milk. J. Diary Sci. 1996;84:1123–1127.

[83] Król J, Litwinczuk Z, Brodziak A, Barlowska J. Lactoferrin, lysozyme and immunoglobulin G content in milk of four breeds of cows managed under intensive production system. Polish J. Vet. Sci. 2010;13:357–361.

[84] Benkerroum N. Antimicrobial activity of lysozyme with special relevance to milk. Afr. J. Biotechnol. 2008;7:4856–4867.

[85] Abd El-Aziz M. Study on lysozyme level, distribution and effect of heat treatment in buffalo and cow milk. Ann. Agric. Sci. (Cairo). 2006;51:439–446.

[86] Montagne P, Cuilliere M, Mole C, Bene M, Faure G. Changes in lactoferrin and lysozyme levels in human milk during the first twelve weeks of lactation. In: Newburg D, editor. Bioactive Components of Human Milk: Springer, USA; 2001. p. 241–247. doi: 10.1007/978-1-4615-1371-1_30

[87] Hussain M, Bakalis S, Gouseti O, Zahoor T, Anjum FM, Shahid M. Dynamic and shear stress rheological properties of guar galactomannans and its hydrolyzed derivatives. Int. J. Biol. Macromol. 2015;72:687–691. doi: 10.1016/j.ijbiomac.2014.09.019

[88] Priyadarshini S, Kansal VK. Purification, characterization, antibacterial activity and N-terminal sequencing of buffalo-milk lysozyme. J. Dairy Res. 2002;69:419–431. doi: 10.1017/S002202990200554X

[89] Persson K, Carlsson Å, Hambleton C, Guidry A. Immunoglobulins, lysozyme and lactoferrin in the teat and udder of the dry cow during endotoxin-induced inflammation. J. Vet. Med. B. 1992;39:165–174. doi: 10.1111/j.1439-0450.1992.tb01154.x

[90] Radwan Y, Elmarimi A. The use of lysoplate test in the diagnosis of subclinical mastitis. Bull. Anim. Health. Prod. Africa. 1987;35:100–135.

[91] Ismail M, Selim S, Soliman R. Changes in lysozyme activity in milk and its significance in the diagnosis of subclinical mastitis in goats. Egypt. Vet. Med. J. 1984;32:59–65.

[92] De Wit J, Van Hooydonk A. Structure, functions and applications of lactoperoxidase in natural antimicrobial systems. Nederlands melk en Zuiveltijdschrift. 1996;50:227–244. doi: 10.1201/9781420039368.ch3

[93] Pruitt KM, Kamau DN. The lactoperoxidase system of bovine and human milk. In: DS Robinson, NAM Eskin, Editors. Oxidative Enzymes in Foods. London: Elsevier Applied Science; 1991. 133–174p

[94] Lorenzen PC, Martin D, Clawin-Rädecker I, Barth K, Knappstein K. Activities of alkaline phosphatase, γ-glutamyltransferase and lactoperoxidase in cow, sheep and goat's milk in relation to heat treatment. Small Ruminant Res. 2010;89:18–23. doi: 10.1016/j.smallrumres.2009.11.013

[95] Reiter B. The lactoperoxidase system of bovine milk. In: Pruitt K, Tenovou J, editors. The lactoperoxidase system: chemistry and biological significance: Marcel Dekker, New York, USA; 1985. p. 123–141. doi: 10.1016/B978-0-08-031739-7.50030-6

[96] Martin-Hernandez M, van Markwijk BW, Vreeman HJ. Isolation and properties of lactoperoxidase from bovine milk. Neth. Milk Dairy J. 1990;44:213–231.

[97] Kussendrager KD, Van Hooijdonk A. Lactoperoxidase: physico-chemical properties, occurrence, mechanism of action and applications. Br. J. Nutr. 2000;84:19–25. doi: 10.1017/S0007114500002208

[98] Gothefors L, Marklund S. Lactoperoxidase activity in human milk and in saliva of newborn infants. Infect. Immun. 1975;11:1210–1215.

[99] Fonteh FA, Grandison AS, Lewis MJ. Factors affecting lactoperoxidase activity. Int. J. Dairy Technol. 2005;58:233–236. doi: 10.1111/j.1471-0307.2005.00227.x

[100] Andrei S, Pintea A, Groza I, Bogdan L, Ciupe S, Matei S. Milk antioxidant enzymes activity in cows with sub clinical mastitis. Scientific Papers – Veterinary Medicine, University of Agricultural Sciences and Veterinary Medicine "Ion Ionescu de la Brad" Iaşi. 2009;52:1–6.

[101] Asadpour R, Tayefi-Nasrabadi H, Moghadam G, Nofouzi K. Correlation between lactoperoxidase activity and somatic cell count for diagnosis subclinical mastitis in early lactation of dairy cows. J. Anim. Vet. Adv. 2008;7:777–779.

[102] Althaus R, Molina M, Rodriguez M, Fernandez N. Analysis time and lactation stage influence on lactoperoxidase system components in dairy ewe milk. J. Dairy Sci. 2001;84:1829–1835. doi: 10.3168/jds.S0022-0302(01)74622-X

[103] McSweeney P, Fox P. Advanced Dairy Chemistry: Basic Aspects: Volume 1A, Proteins. 4th ed. Springer, New York, USA; 2013. doi: 10.1007/978-1-4614-4714-6

[104] Babcock SM, Russell HL. Unorganized ferments of milk: A new factor in the ripening of cheese. Fourteenth annual report of the Wisconsin agricultural experiment station. 1897;161–193.

[105] Johnson A. The composition of milk. In: Webb B, Johnson A, Alford J, editors. Fundamentals of dairy chemistry: AVI Publishing Co. Inc, Westport, CT; 1974. p. 1–57.

[106] He Z, Yu S, Mei G, Zheng M, Wang M, Dai Y, Tang B, Li N. Maternally transmitted milk containing recombinant human catalase provides protection against oxidation for mouse offspring during lactation. Free Radic. Biol. Med. 2008;45:1135–1142. doi: 10.1016/j.freeradbiomed.2008.07.019

[107] Atasever S, Erdem H. An investigation on the determination of mastitis risk levels and milk production traits in Holstein cows. J. Appl. Anim. Res. 2008;34:13–16. doi: 10.1080/09712119.2008.9706932

[108] Atasever S, Erdem H. Association of somatic cell count with catalase enzyme activity in bovine milk. Trends Anim. Vet. Sci. 2010;1:24–27.

[109] Lindmark-Månsson H, Åkesson B. Antioxidative factors in milk. Br. J. Nutr. 2000;84:103–110. doi: 10.1017/S0007114500002324

[110] Yuksel S, Yigit AA, Cinar M, Atmaca N, Onaran Y. Oxidant and antioxidant status of human breast milk during lactation period. Dairy Sci. Technol. 2015;95:295–302. doi: 10.1007/s13594-015-0211-z

[111] Atasever S, Erdem H, Kul E. Relationship between somatic cell count and catalase activity in raw milk of Anatolian buffaloes. Sci. Res. Essays. 2011;6:4109–4112. doi: 10.5897/SRE11.609

[112] Martin HM, Hancock JT, Salisbury V, Harrison R. Role of xanthine oxidoreductase as an antimicrobial agent. Infect. Immun. 2004;72:4933–4939. doi: 10.1128/IAI.72.9.4933-4939.2004

[113] Harrison R. Physiological roles of xanthine oxidoreductase. Drug Metab. Rev. 2004;36:363–375. doi: 10.1081/DMR-120037569

[114] Massey V, Harris C. Milk xanthine oxidoreductase: the first one hundred years. Biochem. Soc. Trans. 1997;25:750–755. doi: 10.1042/bst0250750

[115] Pandya AJ, Khan MMH. Buffalo milk. Handbook of milk of non-bovine mammals. 2006:195–273. doi: 10.1002/9780470999738.ch8

[116] Godber BL, Schwarz G, Mendel RR, Eisenthal R, Harrison R. Molecular characterization of human xanthine oxidoreductase: the enzyme is grossly deficient in molybdenum and substantially deficient in iron-sulphur centres. Biochem. J. 2005;388:501–508. doi: 10.1042/BJ20041984

[117] Atmani D, Benboubetra M, Harrison R. Goats' milk xanthine oxidoreductase is grossly deficient in molybdenum. J. Dairy Res. 2004;71:7–13. doi: 10.1017/S0022029903006514

[118] Benboubetra M, Baghiani A, Atmani D, Harrison R. Physicochemical and kinetic properties of purified sheep's milk xanthine oxidoreductase. J. Dairy Sci. 2004;87:1580–1584. doi: 10.3168/jds.S0022-0302(04)73311-1

[119] Gadave K, Panda S, Singh S, Kalra S, Malakar D, Perugini MA. Structural and functional insights into the catalytic inactivity of the major fraction of buffalo milk xanthine oxidoreductase. PLoS One. 2014;9:1–11. doi: 10.1371/journal.pone.0087618

[120] Baghiani A, Harrison R, Benboubetra M. Purification and partial characterisation of camel milk xanthine oxidoreductase. Arch. Physiol. Biochem. 2003;111:407–414. doi: 10.3109/13813450312331342265

[121] McManaman J, Palmer C, Wright R, Neville M. Functional regulation of xanthine oxidoreductase expression and localization in the mouse mammary gland: evidence of a role in lipid secretion. J. Physiol. 2002;545:567–579. doi: 10.1113/jphysiol.2002.027185

[122] Vorbach C, Scriven A, Capecchi MR. The housekeeping gene xanthine oxidoreductase is necessary for milk fat droplet enveloping and secretion: gene sharing in the lactating mammary gland. Genes Dev. 2002;16:3223–3235. doi: 10.1101/gad.1032702

[123] Battelli MG, Polito L, Bortolotti M, Bolognesi A. Xanthine oxidoreductase-derived reactive species: physiological and pathological effects. Oxid. Med. Cell Longev. 2016:1–8. doi: 10.1155/2016/3527579

[124] Silanikove N, Shapiro F, Leitner G. Posttranslational ruling of xanthine oxidase activity in bovine milk by its substrates. Biochem. Biophys. Res. Commun. 2007;363:561–565. doi: 10.1016/j.bbrc.2007.08.188

[125] Harrison R. Milk xanthine oxidase: properties and physiological roles. Int. Dairy J. 2006;16:546–554. doi: 10.1016/j.idairyj.2005.08.016

Chapter 7

Bacterial Tick-Borne Diseases of Livestock Animals

Şükrü Kirkan, Göksel Erbaş and Uğur Parin

Additional information is available at the end of the chapter

http://dx.doi.org/10.5772/66122

Abstract

Bacterial tick-borne diseases (BTBDs) are very significant in practical one health medicine. In contrast to the restrictions related to diagnostic and clinical application, the control and prevention of bacterial tick-borne diseases are difficult because they require the disruption of a complicated transmission chain, involving vertebrate hosts and ticks, which interact in a constantly changing environment. Q fever, rickettsiosis, borreliosis, ehrlichiosis, anaplasmosis and tularemia are BTBDs, which are discussed in this chapter. Epidemiology, clinical symptoms, diagnosis and prevention subtopics are planning to be prepared under main topics. This chapter presents a brief background of key livestock BTBDs and ticks and reviews the general aspects of BTBDs to identify topics in knowledge and understanding of these diseases, propose areas for future research and draw attention to the need for improved tools for the diagnosis and control of BTBDs.

Keywords: tick, bacterial zoonoses, Q fever, rickettsiosis, borreliosis, ehrlichiosis, anaplasmosis, tularemia

1. Introduction

Bacterial tick-borne diseases (BTBDs) affect the productivity of livestock animals in various regions of the world, leading to a significant adverse impact on the production of resource-poor farming communities. Hence, the livestock industry has become an integral part of world economy, and the large number of dairy cattle is being imported between continents in order to meet an increasing demand of meat and dairy products, it is essential to review current status of bovine BTBDs and to identify diagnosis and prevention in the knowledge of BTBDs and their prevention. Although there has been a recent increase in the number of studies of BTBDs in various geographical regions, information on their prevalence, distribution, tick vectors and control is limited.

2. Bacterial tick-borne diseases of livestock animals

2.1. Q fever

Q fever is a zoonosis associated with *Coxiella burnetii* that is an obligate intracellular parasite classified within the family Rickettsiaceae and which can be divided into six genomic groups based on restriction fragment length polymorphism. Unlike the other members of Rickettsiae, *C. burnetii* is quite resistant to environmental influences and is not dependent upon arthropod vectors for transmission. *C. burnetii* exhibits two antigenic phases: phase I and phase II (**Figure 1**). Phase I organisms are more infectious. The organism has worldwide distribution, although a large serological survey argues that it is not present in New Zealand [1].

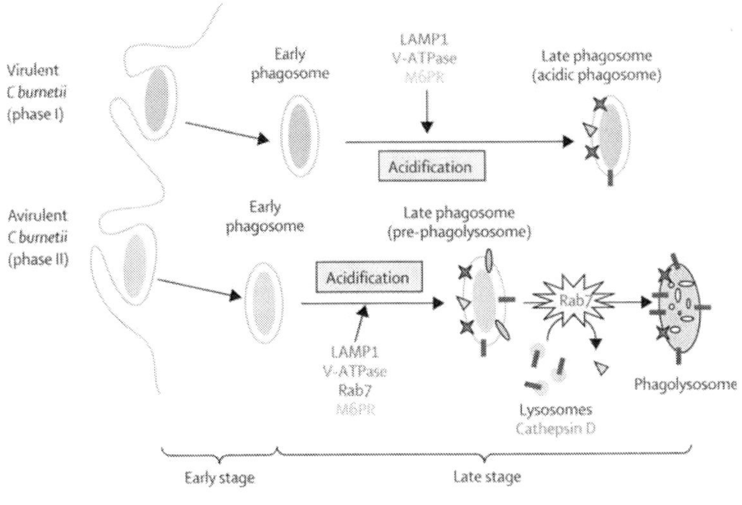

Figure 1. *Coxiella burnetii* mobilization in macrophages [2].

C. burnetii cycles in a wide variety of wildlife species and their ectoparasites. The infection also cycles in domestic animals. Rates of infection in farm animals vary considerably between locations, between countries and with time as there appears to be cycles of infection within regions [3].

In cattle, prevalence figures range from 6 to 82% of cattle and 23 to 96% of herds seropositive depending upon location and country. Seropositivity rates in sheep and goats are similar but also vary according to year and region. There is little information on management or other factors that might influence this variation in prevalence but one study found a significantly higher prevalence in housed cattle compared to cattle at kept at pasture. The transmission of infection is spread by direct contact and inhalation. Infection of non-pregnant animals is clinically silent and is followed by latent infection until pregnancy when there is recrudescence with infection in the intestine, uterus, placenta and udder and excretion from these sites at parturition. The organism is present in high concentration in the placenta and foetal fluids, and subsequent vaginal fluids are also excreted in urine and are present in the faeces of sheep from 11 to 18 days post-partum [4, 5]. Infection can result in abortion, stillbirths or poorly viable lambs but commonly the neonates of infected, excreting, ewes are born clinically normal. Abortion usually does not occur at successive pregnancies but there can be recrudescence of infection and excretion at these pregnancies, especially the one immediately following [6].

Goats also excrete the organism in vaginal discharges for up to 2 weeks, and it is present in goat milk for up to 52 days after kidding and also in faeces. Maximum shedding in cattle also occurs at parturition and for the following 2 weeks but cattle excretes the organism in the milk for at least several months and up to 2 years and infection is common in bulk tank milk [7–10].

There is strain variation in the organism and differences in plasmid sequence types have been correlated with differences in the type of disease occurring in humans. The organism is highly infectious, and it is estimated that the infective dose for humans approximates one organism zoonotic implications in human infection is primarily by inhalation. Sources of infection include such diverse materials such as soil, air-borne dust, wool, bedding and other materials contaminated by urine, faeces or birth products of animals. The potential for human infection from these sources is substantial; for example, ovine manure used as a garden fertilizer has been incriminated as a source. Sheep have traditionally been incriminated as the major reservoir of infection for humans, but the trend for urban populations to locate in close proximity to large dairy herds suggests that cattle could become an increasingly significant reservoir [11–13].

The organism is found in the milk of infected livestock. A significant proportion of seropositive cattle excrete the organism in milk and periods and duration of excretion are variable but may persist at least 2 years. Rates of seropositivity in humans vary markedly between surveys, but there is a higher rate of seropositivity in people (farm workers, veterinarians, livestock dealers, dairy plant and slaughter house workers, shearers, etc.) that are associated with domestic animals and their products and with farm environments [14, 15]. Several incidents of infection in humans have been linked to exposure to parturient sheep and goats [16].

Infection of ruminants can occur at any age and is usually clinically unapparent. In the experimental disease in cattle, anorexia is the only consistent clinical finding. Abortion occurs during the latter part of the lambing period in the flock and in the latter period of pregnancy in individual ewes. The dam shows no signs of impending abortion. As with sheep, infection in goats can be accompanied by abortion, but abortion in cattle is rare although it is recorded. Correlations between herd level seroprevalence and herd fertility are equivocal. There are a number of serological tests available including complement fixation, microagglutination, enzyme-linked immunosorbent assay (ELISA) and indirect immunofluorescence (IF). The IF assay is used as the sero-reference test for the serodiagnosis of Q fever. It can detect antibody to phase variants and can provide epidemiological information as phase I antibody is associated with recent and acute infections and phase II antibody with chronic infections [17].

There are seldom gross lesions in aborted foetuses, but foci of necrosis and inflammation are occasionally seen in the liver, lung and kidney microscopically. The placenta from aborting animals is usually thickened and a purulent exudates or large, red-brown foci of necrosis are typically seen in the thickened intercotyledonary areas. Microscopically, large numbers of necrotic neutrophils are usually visible on the chorionic surface and swollen trophoblasts filled with the organisms can also be found in well-preserved specimens. Examination of placental impression smears stained with Gimenez, Koster's, or other appropriate techniques provides a means of rapid diagnosis. However, care must be taken to avoid confusing Coxiella-infected trophoblasts with cells containing Chlamydophila organisms. Coxiellosis can be confirmed fluorescent antibody staining of fresh tissue or immunohistochemical staining of formalin-fixed samples. In most laboratories, culture is not attempted due to the zoonotic potential of this agent. Polymerase chain reaction (PCR) is the most accurate tool for the diagnosis of infectious abortions. In a previous study, six (4.3%) samples were detected PCR positive out of 138 samples [18]. In another research, *C. burnetii* gene was detected in 34.66% of the samples taken from 200 cattle, 200 sheep and 200 goats in the Aegean region of Turkey [19]. In a multidisciplinary research made with veterinarians, farm workers and butchers, among 92 people, 32 (34.8%) and 9 (9.8%) people were positive and equivocal by ELISA and immunoglobulin G (IgG), respectively. The ELISA positive and equivocal sera were studied further by the immunofluorescence antibody (IFA) test, and seven (7.6%) cases were confirmed with immunoglobulin M (IgM), 39 (42.4%) cases were confirmed with IgG. There was no significant difference for Coxiellosis seropositivity among the profession groups ($p > 0.05$). Only four (4.3%) cases were confirmed with PCR positive [20].

Aborting animals should be isolated for 3 weeks and aborted and placental contaminated material burnt. Ideally, manure should be composted for 6 months before application to fields. Feed areas should be increased to keep them free from contamination with faeces and urine. While Q fever has significant implications for human health, it is not significantly important enough to have generated national or regional control strategies based on control in the animal population [21–23].

Milk and milk products should be pasteurized. Veterinarians dealing with herds that provide raw milk should ensure that these herds are seronegative for *C. burnetii*. Vaccine trials with killed vaccines in animals show a good and persistent antibody response and suggest that vaccination can limit the excretion of the organism. However, there is little economic incentive for a vaccination programme involving livestock, and livestock vaccines are not available in most countries [24].

2.2. Rickettsiosis

The members of the family Rickettsiaceae have cell walls similar to those of other Gram-negative bacteria. Ultra structural studies have shown that the Anaplasmataceae family have outer membranes but lack an obvious peptidoglycan layer [25]. Organisms in the family Rickettsiaceae, referred to as rickettsiae, generally target endothelial cells. Although several new species of rickettsiae have recently been identified in domestic animals using molecular techniques, their pathogenicity is uncertain and currently the only species of veterinary importance in the family Rickettsiaceae is *Rickettsia rickettsii*, the causative agent of Rocky Mountain spotted fever. Many *Rickettsia* species including the causal agents of typhus (*R. prowazekii*), murine typhus (*R. typhi*) and scrub typhus (*R. tsutsugamushi*) are primarily human pathogens. These highly pathogenic organisms have a predilection for the endothelial cells of small blood vessels, resulting in vasculitis and thrombosis in many organs. *Rickettsia* species produce phospholipase that damages the membranes of phagosomes allowing the organisms to escape into the cytoplasm (**Figure 2**). *R. rickettsii* replicates in both the cytoplasm and the nucleus of host cells, inducing cytotoxic effects [26].

Definitive classification of the members of the Rickettsiales is based on 16S ribosomal ribonucleic acid (RNA) sequencing, lipopolysaccharide content and metabolic requirements. In diagnostic laboratories, identification of these organisms is based on the species affected, cell predilection, microscopic appearance and molecular techniques. Some members of the Rickettsiales can be cultured in embryonated eggs or tissue culture cells. These difficult procedures are usually performed only in laboratories engaged in research or vaccine production [28].

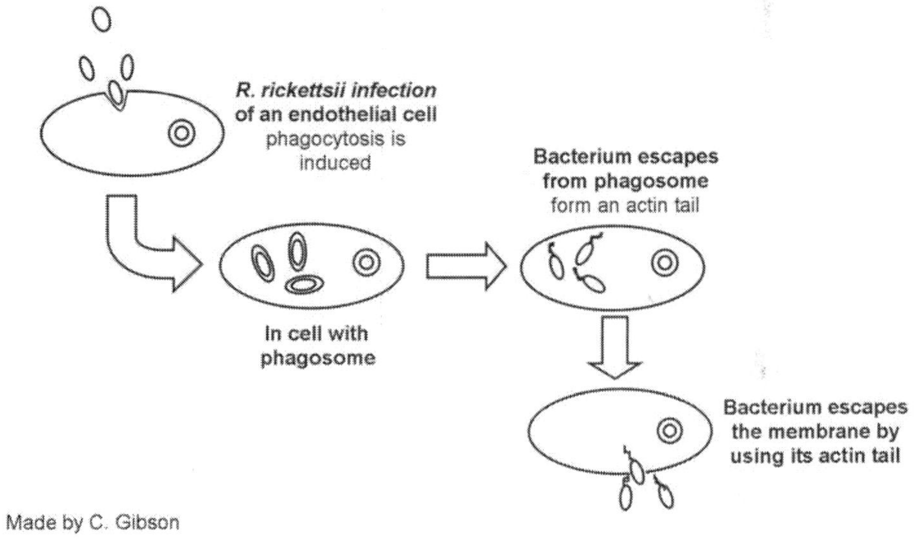

Figure 2. Infection diagram of *R. rickettsii* [27].

R. rickettsii affects mainly humans and dogs. *Rhipicephalus sanguineus* and *Amblyomma cajennense* are the main vectors in Central and South America. Ticks acquire the pathogen while feeding on infected small wild mammals [29].

An infected tick must remain attached for up to 20 hours before salivary transmission to the host occurs. The organisms, which replicate in endothelial cells of infected dogs, produce vasculitis, increased vascular permeability and haemorrhage. Rocky Mountain spotted fever should be considered in dogs with systemic diseases, which have been exposed to ticks in endemic areas. Indirect fluorescent antibody test (FAT) or ELISA demonstrating an increasing antibody titre to *R. rickettsii* is diagnostic. Antibodies are not demonstrable until at least 10 days after infection. A marked thrombocytopenia and leucopoenia may be present during the acute phase of the disease. The disease must be differentiated from acute canine monocytic ehrlichiosis. PCR detection in tick tissues has been described by a number of workers. Tetracycline therapy, which usually produces clinical improvement within 24 hours, must be continued for 2 weeks. Supportive therapy is necessary for severely debilitated dogs. Frequent removal of ticks is recommended. Because the disease is zoonotic, gloves should be worn during this procedure or a forceps should be used [30].

Ticks acquire the pathogen while feeding on infected small wild mammals. *R. rickettsii* is maintained in the tick population by transovarial and transstadial transmission and thus the tick acts as both a reservoir and a vector of the organism. An infected tick must remain attached for up to 20 hours before salivary transmission to the host occurs. The incubation period of the disease is 2–10 days and the course is usually less than 2 weeks. Clinical signs include fever, depression, conjunctivitis, retinal haemorrhages, muscle and joint pain, coughing, dyspnoea and oedema of the extremities [31].

2.3. Borreliosis

Borreliae, which are longer and wider than other spirochaetes, have a similar helical shape. In addition to a linear chromosome, which is unique among bacteria, borreliae possess linear and circular plasmids, some of which appear to be essential for growth and survival of the organism. Although these spirochaetes can cause disease in animals and humans, subclinical infections are also common. Borreliae are transmitted by arthropod vectors. Arthropod vectors are responsible for transmission of *Borrelia* species in animals. Borreliae are obligate parasites in a variety of vertebrate hosts. Although these organisms persist in the environment for short periods, they depend on vertebrate reservoir hosts and arthropod vectors for long-term survival. Associations of certain *Borrelia* species with particular arthropod vectors and reservoir hosts are important in determining the epidemiology of infections with *Borrelia* species. After entering the bloodstream of a susceptible host, borreliae multiply and are disseminated throughout the body (**Figure 3**). Organisms may be demonstrated in joints, brain, nerves, eyes and heart. Whether disease is caused by active infection or by host immune responses to the organism is unclear. Persistent infection leading to the induction of cytokines may contribute to the development of lesions [32]. There may be an association between different genotypes of *Borrelia burgdorferi* and particular clinical syndromes in humans; *B. burgdorferi* sensu stricto (s.s.) is frequently associated with arthritis, *B. garinii* with neurological disease and *B. afzelii* with skin disease [33, 34].

Chickens have been infected experimentally, and it was found that these animals quickly became immune to *B. burgdorferi* s.s. and did not show any clinical symptoms [36]. More recent

http://dx.doi.org/10.5772/66122

studies have shown that pheasants can function as reservoir hosts of *B. garinii* and *B. valaisiana* in the United Kingdom (UK) [37], but no symptoms of disease in infected birds have been reported.

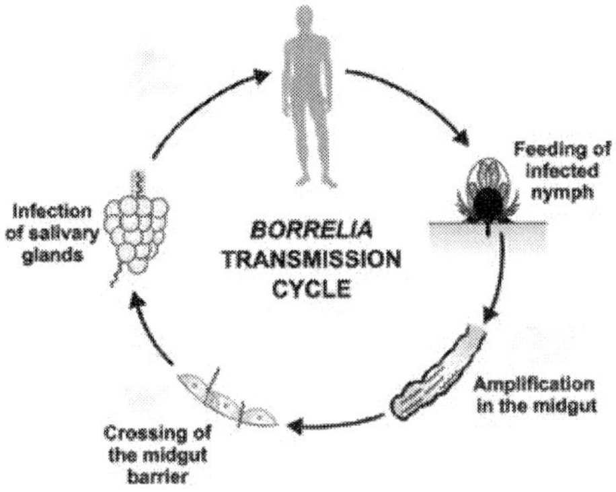

Figure 3. Life cycle of *Borrelia* spp. [35].

Most infections are subclinical. Serological surveys demonstrate that exposure is common in both animal and human populations in endemic areas. The clinical manifestations of Lyme disease are mainly related to the sites of localization of the organisms. Clinical disease is reported frequently in dogs. Symptoms include fever, lethargy, arthritis and evidence of cardiac, renal or neurological disturbance. In the United States of America (USA), arthritis is a common finding whereas neurological disturbance is the most frequent clinical feature in Europe and Japan. The clinical signs in horses are similar to those in dogs and include lameness, uveitis, nephritis, hepatitis and encephalitis. However, some authors observe that definitive evidence of clinical Lyme disease in horses is lacking [38]. Lameness in cattle and sheep associated with *B. burgdorferi* sensu lato infection has been reported.

Laboratory confirmation of Lyme disease may prove difficult because the spirochetes may be present in low numbers in specimens from clinically affected animals. In addition, the organism is fastidious in its cultural requirements. A history of exposure to tick infestation in an endemic area in association with characteristic clinical signs may suggest Lyme disease. Increasing antibody titres to *B. burgdorferi* sensu lato along with typical clinical signs are indicative of disease. Because subclinical infections are common in endemic areas, high titres alone are not confirmatory. The ELISA is extensively used for antibody detection; western immunoblotting is sometimes used for confirmation of ELISA results. It has been shown that ELISA techniques based on this antigen may be able to differentiate naturally infected and vaccinated animals [39]. Immunofluorescence assays may also be used but the results of these methods may be difficult to interpret. Culture of borreliae from clinically affected animals is confirmatory. Cultures in Barbour-Stoenner-Kelly medium should be incubated for 6 weeks under microaerophilic conditions and should be carried out in specialized laboratories. Low numbers of borreliae can be detected in samples by PCR techniques.

Acute Lyme disease responds to treatment with amoxicillin and oxytetracycline. In chronic disease, prolonged or repeated courses of treatment may be required. Acaricidal sprays, baths or dips should be used to control tick infestation. Where feasible, tick habitats such as rough brush and scrub should be cleared. Prompt removal of ticks from companion animals may prevent infection. However, because some tick species can transmit spirochetes shortly after attachment, it cannot be assumed that daily removal of ticks will prevent infection [40].

A number of vaccines, including whole cell bacterins and recombinant subunit vaccines, are commercially available for use in some countries. An outer surface protein A (OspA) recombinant vaccine stimulates the production of antibodies, which are able to kill the borreliae in the gut of the tick and thus prevent infection of the host. However, the benefit of vaccinating animals with currently available vaccines is disputed [41].

2.4. Ehrlichiosis and anaplasmosis

Ehrlichia (Cowdria) ruminantium is a Gram negative, intracellular rickettsial organism in the genus *Ehrlichia*. It occurs in colonies or morulae with a predilection for the vascular endothelium and stains blue with Giemsa stain. The organism is coccoid, 0.2–0.5 μ in diameter. It can now be cultivated *in vitro*, and it can also grow in mice. Cyclical development is believed to take place in intestinal and salivary epithelia of ticks. Although strain differences exist, all isolates possess a major antigenic protein 1 (MAP 1) that is used for diagnosis. However, the antigen cross-reacts with other *Ehrlichia* spp., including *Ehrlichia equi*, the cause of equine granulocytic ehrlichiosis. *Anaplasma* spp. is obligate intraerythrocytic parasites belonging to the order Rickettsiales and infecting ruminants. Infection occurs more sporadically in temperate climate areas. In the USA and other countries, the disease has occurred beyond the boundaries of tick-infested areas and the area distribution in Europe has been advancing northward in recent years with sporadic cases in France, Switzerland, the Netherlands, Hungary and Austria. Anaplasmosis of sheep and goats has a distribution similar to that of cattle. Disease occurs sporadically in the northern states and Canada. In Australia, infection is closely related to the distribution of *Boophilus microplus*, which is restricted to the northern areas. Differences in enzootic and epizootic areas in South America and South Africa are also largely related to tick distribution and climate [42].

Heartwater is limited in its occurrence to sub-Saharan Africa, Madagascar and three Caribbean islands of Guadeloupe, Marie Galante and Antigua. It is one of the main causes of death in imported breeds of cattle, sheep and goats in endemic areas. Heartwater has been diagnosed recently in the island of Mayotte in the Indian Ocean. Measures of disease occurrence in endemic areas, morbidity and mortality rates are low, but the percentage of sera positive titres for heartwater could be as high as 100% in adults, depending on the abundance of tick vectors [43]. Case mortality can be as high as 100% in peracute cases in sheep and goats and as low as 0–10% in cattle. The disease is less severe in indigenous breeds and related game animals reared in enzootic areas, some of which may become symptomless carriers. The N'Dama breed in West Africa is said to be well adapted to heartwater, partly because it can resist tick burdens under the traditional farming system. The method of transmission in the Caribbean, cattle egrets are suspected to spread *Amblyomma variegatum* between islands. Consequently, heartwater is considered threats to the American mainland where potential vectors are present but do not harbour the disease or where the vector may be introduced and become established. Infection in ticks is transmitted transstadially and possibly transovarially. Vertical transmission to calves in colostral milk has also been reported. Several wild ruminants can be infected and become subclinical carriers and reservoirs. Ticks feeding on them can transmit the disease to domestic ruminants. The organism does not infect humans. Cattle are infected with *Amblyomma marginale* and *Amblyomma centrale* and sheep with *Amblyomma avis. A. marginale* will establish in sheep by experimental infection but *A. avis* will not infect cattle. A variety of species of wild ruminants in both North America and Africa can be infected and may have significance as reservoirs for *A. marginale.* In the United States, the black-tail deer in the West Coast region is believed a reservoir and a number of species of antelope play a similar role in South Africa. The prevalence of infection in cattle in endemic areas is very high with seropositivity rates exceeding 60% and often approaching 90%. Seropositivity is much lower in regions that interface between endemic and non-endemic regions. Source and methods of transmission recovery from acute infection result in persistent infection characterized by repetitive cycles of rickettsemia. Persistent carriers are the reservoir for herd infection. The level of parasitemia is often too low for detection by microscopy but can be detected by nucleic acid probe analysis. Transmission occurs biologically by ticks [44].

Heartwater is the most important rickettsial infection of ruminants in Africa and it is regarded as the most important disease of ruminants. In general, heartwater is a more serious problem where *Amblyomma habraeum* is the vector. In countries or regions where there is endemic stability, losses from heartwater are minimal until new animals are introduced. On the other hand, since most losses are in exotic animals, heartwater is a major constraint to livestock improvement in sub-Saharan Africa. Furthermore, it has the potential to spread from the Caribbean to the American mainland. Heartwater requires the vector tick to get established in any community. Therefore, there is concern about possible illegal importation of infected animals or ticks to southern United States where potential vectors exist. In ewes intra-uterine infection appears to occur with ease in experimental cases provided the ewe is exposed during the latter two-thirds of pregnancy. In sheep and goats, infection is usually subclinical but in some cases, particularly in goats, a severe anaemia may occur and a clinical picture similar to that found in cattle may be seen. Severe reactions of this type in goats are most frequent when the animals are suffering from concurrent disease. Goats may show hyper excitability and may bite at inanimate objects. The experimental disease in lambs includes fever, constipation or diarrhoea, pale, icteric conjunctivae and severe anaemia 15–20 days after inoculation. The anaemia is not completely resolved in 3–4 months. *A. avis* are usually situated at the periphery of erythrocytes but as many as 40% of infested cells may show sub-marginal protozoa [45].

The incubation period is 1–3 weeks after transmission in tick saliva. Depending on the susceptibility of individual animals and the virulence of the infecting organism, the resulting disease may be peracute, acute, subacute or mild and unapparent. Peracute cases show only high fever and death with terminal convulsions in 1–2 days. Acute cases are more common and have a course of about 6 days. A sudden febrile reaction is followed by inappetence and rapid breathing followed by the classical nervous syndrome that is characteristic of heartwater.

It comprises ataxia, chewing movements. Profuse, fetid diarrhoea is frequent. Subacute cases are less severe but may terminate in death in 2 weeks or the animal may gradually recover. The mild form is often subclinical and is seen mainly in indigenous animals and wild ruminants with high natural or induced resistance. The case mortality rate in peracute cases is 100%, in acute cases 50–90% and in calves below 4 weeks of age it is 5–10%, most animals recover in mild cases [46].

Haematological changes in heartwater are not specific but there may be thrombocytopenia, neutropenia, eosinopenia and lymphocytosis. Confirmatory diagnosis is based on identifying the *Rickettsia* in capillary endothelial cells using a Giemsa stained squash preparation of brain tissue at post-mortem. The rickettsiae occur as blue to reddish-purple colonies or morulae of five to several hundred coccoid organisms (0.2–0.5 μ in diameter) in the cytoplasm of the cells. An immunohistochemical staining technique has also been described [47]. Injection of blood into sheep may also be used as a diagnostic procedure. The available serological test is an indirect fluorescent antibody test used for surveys but the close antigenic relationship with other *Ehrlichia* spp. often leads to false positives. An ELISA based on recombinant MAP 1 protein of *C. ruminantium* was reported to be more sensitive. In general, clinical detection of heartwater is not always easy because all serological assays so far available have poor sensitivity or specificity. Diff-Quik staining of blood smears is as accurate as Giemsa in the detection of *A. marginale* and can be completed in 15 seconds as compared to nearly an hour for Giemsa. There are no diagnostic clinical chemistry findings. A rapid card agglutination test, which tests serum or plasma for antibodies against *A. marginale*, is cheap and quick, and sufficiently accurate to be used as a herd test. Currently, in most countries, the card agglutination and complement fixation (CF) tests are routinely available. It is also an accurate test for selecting recently affected animals. A dot-ELISA with high sensitivity, specificity and predictive value is also described and could be particularly applicable to field examinations. A competitive inhibition ELISA test, with high sensitivity and specificity, has been developed that detects antibody to a major surface protein that is conserved among *Anaplasma* species; this test can be used to detect cattle persistently infected for as long as 6 years. Vaccinated animals may react to all of the serological tests for periods of over 1 year. Nucleic probe analysis can be used to detect low levels of parasitaemia. Transmission to splenectomised animals has been used to detect carriers but is expensive and is now replaced by PCR in countries where this technology is available [48]. A polymerase chain reaction assay has therefore been suggested as the method of choice for detection of *E. ruminantium* infection [49].

Field cases of heartwater are difficult to treat successfully because available drugs are effective only in early febrile stages before neurological signs develop. In the early stages, short-acting tetracyclines at 10–20 mg/kg body weight (BW) and long-acting forms at reduced doses are effective. Sulphonamides can also be used in the early stages but are less effective. Hyperimmune serum is said to be of no curative value. Supportive therapy to reduce either the pulmonary oedema or the neurologic signs or to stabilize membranes in general is being investigated but with little success. Chemoprophylaxis involves administration of tetracyclines or subcutaneous implantation of doxycycline in susceptible animals when they are introduced into an endemic area. Results are not always predictable. Anaplasmosis treatment is with tetracyclines. Treatment of clinical disease can be with oxytetracycline, 6–10 mg/kg BW daily for 3 days, or a one dose application of long-lasting 20 mg/kg oxytetracycline intramuscularly. The convalescent period is long. Concurrent administration of estradiol cypionate (14.3 mg/kg BW intramuscularly) appears to improve the rate of recovery by promoting parasitemia during treatment. Tetracycline treatment will not eliminate infection and immunity will persist. Blood transfusions are indicated in animals with a packed cell volume (PCV) less than 15%. Rough handling must be avoided. Imidocarb (3 mg/kg BW) is also an effective treatment for clinical cases and does not interfere with the development of acquired immunity to *A. marginale*. The risk for infection in the rest of the herd should be assessed and, if necessary, temporary or prolonged protection should be provided. Protection can be provided by tetracyclines, or by vaccination [50].

Past efforts to control heartwater were based on intensive acaricide treatment in endemic areas. It involved frequent use of acaricides (plunge dipping) up to 52 times a year. This has now been shown to be environmentally unfriendly, economically unsustainable, and would invariably lead to animals that remained always susceptible. For example, it was observed in Zimbabwe that large farms applying acaricides very frequently (more than 30 times per annum) had higher morbidity and mortality than farms applying acaricides less frequently. Vaccination is based on infection and treatment regimen that was first developed more than 50 years ago. It involves an intravenous injection of virulent organisms in cryopreserved sheep blood, followed by treatment with tetracyclines at the first indication of fever. Most control programmes in enzootic areas are based on increasing the resistance of the population by immunization. In any vaccination programme, particular attention should be paid to the animals at high risk, particularly animals brought in from non-enzootic areas, those in surrounding similar areas to which infection may be spread by expansion of the vector population under the influence of suitable climatic conditions, and animals within the area are likely to be exposed to climatic or nutritional stress [51].

Vaccination may lead to some deaths, the immunity may wane in the absence of reinfection, and animals may become carriers. More recently, cattle were successfully immunized for up to 10 months with a killed vaccine from a lysate of *E. ruminantium* formulated in Freund's adjuvant. In another study, the use of inactivated vaccines from cell-cultured *E. ruminantium* combined with an adjuvant led to a reduction in mortality from heartwater in cattle, sheep and goats exposed to field challenges in Botswana, Zambia, Zimbabwe, and South Africa. Experimental studies using deoxyribonucleic acid (DNA) recombinant vaccines so far have met with only limited success. Killed *A. marginale* are usually in an adjuvant vehicle. The vaccine requires two doses, 4 weeks apart, the last dose given at least 2 weeks before the vector season. However, there is a risk for neonatal isoerythrolysis. This can be reduced by vaccinating only empty cows and avoiding unnecessary booster injections. When this vaccine is used in the face of an outbreak, tetracyclines can also be given to provide temporary protection during the period of development of immunity; tetracyclines do not interfere with the development of this immunity. Preliminary reports of the efficacy of DNA vaccines are not encouraging. A living *A. centrale* vaccine is used extensively in Australia, Africa, Israel and Latin America, but not in the USA and there is some reluctance to introduce it into areas where *A. centrale* does not already occur. A single vaccination is used in endemic areas and the immunity is reinforced by continuous challenge and considered to persist for life in tick areas. Vaccine administration is limited to the relatively resistant age group below 1 year of age, to the winter months when vectors are sufficiently rare to avoid the chance of spread to other age groups, and to circumstances where animals that react severely can be restrained and treated adequately. The method has the serious disadvantage of creating a large population of carrier animals which may subsequently spread the disease. Attenuated vaccines have been attempted by irradiation of strains and passage of the organism through sheep or deer and the use of naturally low virulence isolates [52, 53].

For tick control, flumenthrin 1% pour on at 45 days interval was found to provide effective protection of Friesian/Zebu crossbred cattle against important ticks, but it must be applied correctly at the recommended dose. Pure Zebu and N'Dama cattle would probably require less frequent applications, Flumenthrin pour-on is gradually replacing plunge dipping for the control of ticks and tick-borne diseases in general. Other than routine surveillance, there are no special biosecurity concerns with heartwater, since transmission requires the presence of the vector [54].

2.5. Tularemia

The disease causes acute septicaemia, with localization and granulomatous lesions and the organs (particularly the liver and spleen). Signs are very non-specific, as expected with bacteraemia, and include fever, anorexia, lethargy, and in some cases cough, rapid respiration or diarrhoea. Stiffness and oedema of the limbs may be seen. The incubation period of the disease is usually 2–14 days in companion animals [55].

Tularemia is a highly contagious disease occurring principally in wild animals but it may transmit to farm animals, causing septicaemia and high mortality. *Francisella tularensis* is the causative organism [56].

Tularemia is primarily restricted in its occurrence to countries in the northern hemisphere and occurs in most of them. In North America, the disease is most prevalent in farm animals in the north-western states of the USA and the adjoining areas of Canada, although in these areas it is rare and the majority of reports in livestock are historical. *F. tularensis* has a wide host range and is recorded in over 100 species of bird and wild and domestic animal. Disease is recorded among farm animals, most commonly in sheep and pigs and to a lesser extent in calves, which appear more resistant but can be infected in association with heavy tick infestation [57]. Sheep and pigs of all ages are susceptible but most losses occur in lambs, and in pigs clinical illness occurs only in piglets. There is a sharp seasonal incidence, the bulk of cases occurring during the spring months. The morbidity rate in affected flocks of sheep is usually about 20% but may be as high as 40%, and the mortality rate may reach 50%, especially in young animals. With sheep, transmission occurs chiefly by the bites of the wood tick, *Dermacentor andersoni*, and from *Haemaphysalis otophila*, the ticks becoming infected in the early part of their life cycle when they feed on rodents. In Europe *Ixodes ricinus* and *Dermacentor reticulatus* are vectors [58]. Transstadial and transovarial transmission occurs in the tick. The adult ticks infest sheep, and pastures bearing low shrubs and brush are particularly favourable to infestation. The ticks are

http://dx.doi.org/10.5772/66122

found in greatest numbers on the sheep around the base of the ears, the top of the neck, the throat, axillae and udder. It is assumed that sheep are relatively resistant to tularemia but become clinically affected when the infection is massive and continuous. Transmission to pigs and horses is thought to occur chiefly by tick bites but mechanical transmission to laboratory animals does occur with tabanid and blackflies. Tularemia is an acute septicaemia but localization occurs, mainly in the parenchymatous organs, with the production of granulomatous lesions [59].

In the sheep, the incubation period has not been determined. A heavy tick infestation is usually evident. The onset of the disease is slow with a gradually increasing stiffness of gait, dorsiflexion of the head and a hunching of the hindquarters; affected animals lag behind the group. The pulse and respiratory rates are increased, the temperature is elevated up to 42°C (107°F), and a cough may develop. There is diarrhoea, the faeces being dark and fetid, and urination occurs frequently with the passage of small amounts of urine. Body weight is lost rapidly, and progressive weakness and recumbency develop after several days, but there is no evidence of paralysis, the animal continuing to struggle while down [60]. Death occurs usually within a few days but a fatal course may be as long as 2 weeks. Animals that recover commonly shed part or the entire fleece but are solidly immune for long periods. In pigs, the disease is latent in adult pigs but young piglets show fever up to 42°C, accompanied by depression, profuse sweating and dyspnoea. The incubation period of the disease is about 7–10 days. In horses, fever (up to 42°C) and stiffness and oedema of the limbs occur. Foals are more seriously affected and may show dyspnoea and incoordination in addition to the above signs [61]. Necropsy usually reveals ticks on the carcass. Often, reddened or necrotic areas appear in and under the skin at the site of the infected bites. Regional lymph nodes may be swollen and congested. Congestion and oedema of the lungs are common [62].

An agglutination test is available for the diagnosis of tularemia, a titre of 1:50 being regarded as a positive test in pigs. Serum from pigs affected with brucellosis does not agglutinate tularemia antigen, but serum from pigs affected with tularemia agglutinates brucellosis antigen. Cross-agglutination between *F. tularensis* and *Brucella abortus* is less common in sheep and an accurate diagnosis can be made on serological grounds because of the much greater agglutination that occurs with the homologous organism. Titres of agglutinins in affected sheep range from 1:640 to 1:5000 and may persist at levels of 1:320 for up to 7 months. A titre of 1:200 is considered as positive in sheep. In horses the titres revert to normal levels in 14–21 days. An intradermal sensitivity test using 'tularin' has been suggested as being more reliable as a diagnostic aid in pigs than the agglutination test, but is unreliable in sheep. In sheep, large numbers of ticks may be present on the hides of fresh carcasses. In animals that have been dead for some time, dark red subcutaneous areas of congestion up to 3 cm in diameter are found and may be accompanied by local swelling or necrosis of tissues [63]. These lesions mark the attachment sites of ticks. Enlargement and congestion of the lymph nodes draining the sites of heaviest tick infestation are often noted. Pulmonary oedema, congestion or consolidations are inconstant findings. In pigs, the characteristic lesions are pleuritis, pneumonia and abscessation of submaxillary and parotid lymph nodes. The organisms can be isolated from the lymph nodes and spleen, and from infected ticks. Isolation can also be effected by experimental transmission to guinea pigs. Techniques such as immunoperoxidase staining of fixed specimens and PCR of fresh tissues can circumvent the need for culture of this zoonotic agent. Samples for confirmation of diagnosis are based on

- Bacteriology: lung, lymph node, spleen (CULT—requires cysteine-enriched media, PCR).
- Histology: above tissues plus liver, fixed in formalin [64].

Treatment early in the course of infection is effective. Aminoglycosides, tetracyclines or cephalosporins all are probably beneficial initially, until results of antimicrobial susceptibility testing are available. Streptomycin, gentamicin, the tetracyclines and chloramphenicol are effective treatments in humans and companion animals. Oxytetracycline (6–10 mg/kg BW) has been highly effective in the treatment of lambs and much more effective than penicillin and streptomycin. Insecticide removal of ticks from affected animals and herdmates is important. An outbreak of tularemia in sheep can be rapidly halted by spraying or dipping with insecticide to kill the vector ticks. In areas where ticks are enzootic, sheep should be kept away from shrubby, infested pasture or sprayed regularly during the months when the tick population is greatest. An experimental live attenuated vaccine has been developed, but there is no routine vaccination of livestock [65].

3. Conclusion

Given that the livestock industry has become an integral part of world economy and a large number of dairy cattle are being imported between countries, in order to meet an increasing demand of meat and dairy products, it is essential to review current status of bovine BTBDs and to identify diagnosis and prevention in the knowledge of BTBDs and their control. Although there has been a recent increase in the number of studies of BTBDs in various regions and facilities, information on their prevalence, distribution, tick vectors and control is limited. This chapter provides a brief background on key bovine BTBDs and ticks and reviews the general aspects of bovine BTBDs to identify gaps in knowledge and understanding of these diseases, propose areas for future research and draw attention to the need for improved tools for the diagnosis and control of BTBDs.

Author details

Şükrü Kirkan*, Göksel Erbaş and Uğur Parin

*Address all correspondence to: skirkan@adu.edu.tr

Department of Microbiology, Faculty of Veterinary Medicine, Adnan Menderes University, Aydin, Turkey

References

[1] Maurin M, Raoult D. Q fever. Clin. Microbiol. Rev. 1999;12:518–553.

[2] Raoult D, Marrie TJ, Mege JL. Natural history and pathophysiology of Q fever. Lancet Infect. Dis. 2005;5:219–26.

[3] Mertens K, Samuel JE. Bacteriology of Coxiella. In: Raoult D, Parola P, editors. Rickettsial Diseases. New York: Informa Healthcare Inc.; 2007. pp. 257–270.

[4] Marrie TJ, Raoult D. Update on Q fever, including Q fever endocarditis. Curr. Clin. Top. Infect. Dis. 2002;22:97–124.

[5] Stein A, Raoult D. Pigeon pneumonia in Provence. A bird borne Q fever outbreak. Clin. Infect. Dis. 1999;29:617–620.

[6] Stoker MG, Marmion BP. The spread of Q fever from animals to man. The natural history of a rickettsial disease. Bull WHO. 1955;13:781–806.

[7] Marrie TJ, Langille D, Papukna V, Yates L. Truckin pneumonia-an outbreak of Q fever in a truck repair plant probably due to aerosols from clothing contaminated by contact with newborn kittens. Epidemiol. Infect. 1989;102:119–127.

[8] Babudieri B. Q fever. A zoonosis. Adv. Vet. Sci. 1959;5:82–182.

[9] Sanford ES, Josephson GKA, MacDonald A. *Coxiella burnetii* (Q fever) abortion storms in goat herds after attendance at an annual fair. Can. Vet. J. 1994;35:376–378.

[10] Huebner RJ, Bell JA. Q fever studies in Southern California. Summary of current results and a discussion of possible control measures. J. Am. Med. Assoc. 1951;145:301–305.

[11] Webster JP, LLoyd G, Macdonald DW. Q fever (*Coxiella burnetii*) reservoir in wild brown rat (Rattus norvegicus) populations in the UK. Parasitology. 1995;110:31–35.

[12] Lang GH. Serosurvey of *Coxiella burnetii* infection in dairy goat herds in Ontario. Can. J. Vet. Res. 1988;52:37–41.

[13] Marrie TJ. Epidemiology of Q fever. In: Raoult D, Parola P, editors. Rickettsial Diseases. New York: Informa Healthcare Inc.; 2007. pp. 281–289.

[14] Arricau-Bouvery N, Rodolakis A. Is Q fever an emerging or reemerging zoonosis? Vet. Res. 2005;36:327–349.

[15] Dupuis G, Peter O, Luthy R, Nicolet J, Peacock M, Burgdorfer W. Serological diagnosis of Q fever endocarditis. Eur. Heart J. 1986;7:1062–1066.

[16] Peacock MG, Philip RN, Williams JC, Faulkner RS. Serological evaluation of Q fever in humans: enhanced phase I titers of immunoglobulins G and A are diagnostic for Q fever endocarditis. Infect. Immun. 1983;41:1089–1098.

[17] Sidi-Boumedine K, Rousset E, Henning K, Ziller M, Niemczuck K, Roest HIJ, Thiéry R. Development of harmonised schemes for the monitoring and reporting of Q-fever in

animals in the European Union. EFSA Scientific Report on Question No EFSA-Q-2009-00511. 2010. 48 p.

[18] Kırkan Ş, Kaya O, Tekbıyık S, Parın U. Detection of *Coxiella burnetii* in cattle by using PCR. Turk. J. Vet. Anim. Sci. 2008;32:215–220.

[19] Parın U, Kaya O. Detection of *Coxiella burnetii* prevalence in bovine, ovine and caprine herds. Ankara Üniv. Vet. Fak. Derg. 2015;62:177–181.

[20] Eyigör M, Kırkan Ş, Gültekin B, Yaman S, Tekbıyık S, Aydın N. Detection of antibodies against *Coxiella burnetii* in risk groups for Q fever: a study with ELISA and IFA tests. Turk. J. Inf. 2006;20:31–36.

[21] Chiu CK, Durrheim DN. A review of the efficacy of human Q fever vaccine registered in Australia. N.S.W. Public Health Bull. 2007;18:133–136.

[22] Kovacova E, Kazar J. Q fever-still a query and underestimated infectious disease. Acta Virol. 2002;46:193–210.

[23] Waag DM, England MJ, Tammariello RF. Comparative efficacy and immunogenicity of Q fever chloroform:methanol residue (CMR) and phase I cellular (Q-Vax) vaccines in cynomolgus monkeys challenged by aerosol. Vaccine. 2002;20:2623–2634.

[24] OIE. Terrestrial Manual [Internet]. 2015. Available from: http://www.oie.int/fileadmin/Home/eng/Health_standards/tahm/2.01.16_Q_FEVER.pdf [Accessed: 2016-07-25]

[25] Cowan GO. Rickettsial infections. In: Cook GC, Zumla A, editors. Manson's Tropical Diseases. 21st ed. London Saunders: Elsevier Science, Health Sciences Division; 2003, Vol. 50, pp. 891–906.

[26] Walker DH, Raoult D. *Rickettsia rickettsii* and other spotted fever group Rickettsiae (Rocky Mountain spotted fever and other spotted fevers). In: Mandell GL, Bennet JE, Doalin R, editors. Principles and Practice of Infectious Diseases. Philadelphia: Churchill Livingstone; 2000. pp. 2035–2042.

[27] Felton B, Lacroix Z, Denisson A, Jacques J. Rocky Mountain Spotted Fever [Internet]. 2012. Available from: https://bacterialandviraldiseases.wikispaces.com [Accessed: 2016-07-03]

[28] Anacker RL, McDonald GA, List RH, Mann RE. Neutralizing activity of monoclonal antibodies to heat-sensitive and heat-resistant epitopes of Rickettsia rickettsii surface proteins. Infect. Immun. 1987;55:825–827.

[29] McDonald GA, Anacker RL, Mann RE Milch LJ. Protection of guinea pigs from experimental Rocky Mountain spotted fever with a cloned antigen of Rickettsia rickettsii. J. Infect. Dis. 1988;158:228–231.

[30] Vishwanath S, McDonald GA, Watkins NG. A recombinant Rickettsia conorii vaccine protects guinea pigs from experimental boutonneuse fever and Rocky Mountain spotted fever. Infect. Immun. 1990;58:646–653.

[31] Diaz-Montero CM, Feng HM, Crocquet-Valdes PA, Walker DH. Identification of protective components of two major outer membrane proteins of spotted fever group Rickettsiae. Am. J. Trop. Med. Hyg. 2001;65:371–378.

[32] Roberts ED, Bohn RP, Lowrie RC. Pathogenesis of Lyme neuroborreliosis in the Rhesus monkey: the early disseminated and chronic phases of disease in the peripheral nervous system. J. Infect. Dis. 1998;178:722–732.

[33] Van Dam AP, Kuiper H, Vos K. Different genospecies of Borrelia burgdorferi are associated with distinct clinical manifestations of Lyme borreliosis. Clin. Infect. Dis. 1993;17:708–717.

[34] Gray JS. Tick-borne pathogen interactions. Infect. Dis. Rev. 1999;1:117–121.

[35] Hajdušek O, ŠímaR, Ayllón N, Jalovecká M, Jan Perner J, de la Fuente J, Kopáček P. Interaction of the tick immune system with transmitted pathogens Front. Cell. Infect. Microbiol. 2013;26:1–15.

[36] Piesman J, Dolan MC, Schriefer ME, Burkot TR. Ability of experimentally infected chickens to infect ticks with the Lyme disease spirochete, Borrelia burgdorferi. Am. J. Trop. Med. Hyg. 1996;54:294–298.

[37] Kurtenbach K, Peacey M, Rijpkema SGT, Hoodless AN, Randolph SE Nuttall PA. Differential transmission of genospecies of Borrelia burgdorferi sensu lato by game birds and small rodents in England. App. Environ. Microbiol. 1998;64:1169–1174.

[38] Butler CM, Houwers DJ, Jongejan F, van der Kolk JH. Borrelia burgdorferi infections with special reference to horses. A review. Vet. Quart. 2005;27:146–156.

[39] O'Connor TP, Esty KJ, Hanscom JL, Shields P, Philipp MT. Dogs vaccinated with common Lyme disease vaccines do not respond to IR6, the conserved immunodominant region of the VlsE surface protein of Borrelia burgdorferi. Clin. Diag. Lab. Immunol. 2004;11:458–462.

[40] Korenberg EI, Moskvitina GG. Interrelationships between different Borrelia genospecies and their principal vectors. J. Vector Ecol. 1996;21:178–185.

[41] Littman MP, Goldstein RE, Labato MA, Lappin MR, Moore GE. ACVIM small animal consensus statement on Lyme disease in dogs: diagnosis, treatment, and prevention. J. Vet. Int. Med. 2006;20:422–434.

[42] Anonymous [Internet]. 2013. Ehrlichiosis and Anaplasmosis: Zoonotic Species Available from: http://www.cfsph.iastate.edu/Factsheets/pdfs/ehrlichiosis.pdf [Accessed: 2016-08-12]

[43] Parola P, Raoult D. Ticks and tickborne bacterial diseases in humans: an emerging infectious threat. Clin Infect Dis 2001;32:897–928. Erratum: Clin. Inf. Dis. 33:749.

[44] Raoult D, Roux V. Rickettsioses as paradigms of new or emerging infectious diseases. Clin. Microbiol. Rev. 1997;10:694–719.

[45] Anonymous [Internet]. 2012. Available from: http://www.bada-uk.org [Accessed: 2016-08-13]

[46] Cunha B, Domachowske J, Keim SM, Abuhammour W, Bennett NJ. Ehrlichiosis [Internet]. 2012. Available from: http://emedicine.medscape.com/infectious_diseases [Accessed: 2016-06-13]

[47] Dumler JS, Barbet AF, Bekker CP, Dasch GA, Palmer GH, Ray SC, Rikihisa Y, Rurangirwa FR. Reorganization of genera in the Families Rickettsiaceae and Anaplasmataceae in the order *Rickettsiales:* unification of some species of *Ehrlichia* with *Anaplasma, Cowdria* with *Ehrlichia*, and *Ehrlichia* with *Neorickettsia*, descriptions of five new species combinations and designation of *Ehrlichia equi* and "HGE agent" as subjective synonyms of *Ehrlichia phagocytophila*. Int. J. Syst. Evol. Microbiol. 2001;51:2145–2165.

[48] OIE, Terrestrial Manual. [Internet]. 2015. Available from: http://www.oie.int/fileadmin/Home/eng/Health_standards/tahm/2.04.01_BOVINE_ANAPLASMOSIS.pdf [Accessed: 2016-06-21]

[49] Alberti A, Zobba R, Chessa B, Addis MF, Sparagano O, Pinna Parpaglia ML, Cubeddu T, Pintori G, Pittau M. Equine and canine Anaplasma phagocytophilum strains isolated on the island of Sardinia are phylogenetically related to pathogenic strains from the United States. Appl. Environ. Microbiol. 2005;71:6418–6422.

[50] Uilenberg G. Note sur les babésioses et l'anaplasmose des bovins à Madagascar. IV. Note additionelle sur la transmission. Rev. Élev. Méd. Vét. Pays Trop. 1970;23:309–312.

[51] Connell M, Hall WTK. Transmission of *Anaplasma marginale* by the cattle tick *Boophilus microplus*. Aust. Vet. J. 1972;48:477.

[52] Leatch G. Preliminary studies on the transmission of *Anaplasma marginale* by *Boophilus microplus*. Aust. Vet. J. 1973;49:16–19.

[53] Anonymous, Tick-borne Livestock Diseases and Their Vectors. [Internet]. 2016. Available from: http://www.fao.org/docrep/004/x6538e/X6538E02.htm [Accessed: 2016-06-23]

[54] MERCK, Anaplasmosis. [Internet]. 2016. Available from: http://www.merckvetmanual.com/mvm/circulatory_system/blood_parasites/anaplasmosis.Html [Accessed: 2016-06-24]

[55] Mchardy N. Immunization against anaplasmosis: a review. Prev. Vet. Med. 1984;2:135–146.

[56] Kocan KM, DE LA Fuente J, Guglielmone AA, Melendéz RD. Antigens and alternatives for control of *Anaplasma marginale* infection in cattle. Clin. Microbiol. Rev. 2003;16:698–712.

[57] Ellis J, Oyston PCF, Green M, Titball RW. Tularemia, Clin. Microbiol. Rev. 2002:15;631–646.

http://dx.doi.org/10.5772/66122

[58] OIE, Terrestrial Manual. [Internet]. 2008. Available from: http://web.oie.int/fr/normes/mmanual/2008/pdf/2.01.18_TULAREMIA.pdf [Accessed: 2016-06-29]

[59] MERCK, Overview of Tularemia. [Internet]. 2016. Available from: http://www.merckvetmanual.com/mvm/generalized_conditions/tularemia/overview_of_tularemia.html [Accessed: 2016-06-30]

[60] Bell JF. Tularaemia. In: Steel JH, editor. CRC Handbook Series in Zoonoses. Section A: Bacterial, Rickettsial, and Mycotic Diseases. 2nd ed. Florida: CRC Press; 1980. pp. 161–193.

[61] Morner T, Addison E. Tularemia. In: Williams ES, Barker IK, editors. Infectious Diseases of Wild Mammals. 3rd ed. Iowa: Iowa State University Press; 2001. pp. 303–313.

[62] Smith BP. Large Animal Internal Medicine. 4th ed. Missouri: Elsevier Health Sciences; 2008. 1184 p.

[63] AVMA, Tularemia. [Internet]. 2016. Available from: https://www.avma.org/News/Journals/Collections/Documents/javma_222_6_725.pdf [Accessed: 2016-07-02]

[64] Anonymous. Tularemia, Kosovo. Wkly. Epidemiol. Rec. 2000;75:133–134.

[65] Gilligan PH. *Pseudomonas* and *Burkholderia*. In: Murray PR, Baron EJ, Pfaller MA, Tenover FC, Yolken RH, editors. 6th ed. Manual of Clinical Microbiology. Washington DC: American Society for Microbiology Press; 1995. pp. 509–519.

Chapter 8

Contaminants in Animal Products

Emine Baydan, Murat Kanbur, Emre Arslanbaş,
Farah Gönül Aydın, Semra Gürbüz and
Muhammed Yasin Tekeli

Additional information is available at the end of the chapter

http://dx.doi.org/10.5772/67064

Abstract

Organic and conventional animal products may include residues of veterinary drugs and environmental contaminant. Food contaminants can cause consumer illness such as allergy, immunosuppression, cancer, teratogenicity, mutagenicity and genotoxicity. Therefore, their control is an important issue in terms of public health. In this article, information is given about contaminants such as bacterial, fungal, metal pesticides and veterinary drug that can be found in organic and conventional animal products. In addition, the effects of various cooking and freezing processes on contaminants in animal foods and their legal regulation have been mentioned.

Keywords: contaminant, animal, product, organic, conventional

1. Introduction

Chemical substances have been used excessively in order to increase the agricultural productivity since the 1940s. Applications initially led to an apparent increase in yield. However, in the later process, effectiveness of these substances decreased due to the development of resistance against chemicals, particularly used in combating agricultural pests, thus this situation resulted in either excessive use of them to obtain better response or development of new drugs with high expenses. Moreover, in this course, human, animal and environmental health problems are reached much more serious extent besides the economic losses [1]. This situation, particularly, in the developed countries has led people to consume more safe products. The current approach is more comprehensive, which ensure the dissemination of sustainable practices in every production area in order to leave a healthier world for future generations.

Organic farming, emerged in this context, is accepted as the farming not allowing the use of any of the substances/applications such as growth promoters, antibiotics, genetically modified organisms and irradiation, which are considered to be harmful to human health, and providing safer foods concerning nitrates, pesticides and harmful elements (heavy metals, particularly cadmium) and rich in phenolic compounds and vitamins [2, 3].

2. Contaminants in organic animal products

Organic farming refers to breeding systems that do not use chemical inputs in which the priority is given to animal welfare and quality of healthy products [4]. In organic livestock production system, vaccination is subjected to conditional permission [5]. Organic farming has increased intensely for the last 10 years in Europe. However, difficulties in the treatment of animal diseases due to failure in achieving the standards of organic farming has led to insufficient development of organic farming and to have a small share in the overall agriculture [4]. Although milk is the most commonly produced products among the organic animal products, its production amount is still considerably lower than that produced by the conventional method. The organic meat production has been recently introduced; therefore, it is difficult to find certified breeders [6]. According to 2001 data, concerning the organic animal breeding, Europe takes the first place with 57.9%, which is followed by the North (15.5%) and South America (13.9%). Organic animal product quality varies depending on various factors such as animal species and diet types. Although, concerning some parameters, organic animal products are superior to conventional animal products, generally, they are considered not to be superior to conventional ones in terms of quality [7]. Despite all this, the organic products are generally regarded as excellent products. For this reason, researches on the contamination in organic products, especially, organic animal products are limited [8]. However, unlike the conventional farming, lack of the use of protective products in organic products can lead to early deterioration of a product, to the risk for mold formation and to the emergence of harmful pathogens. On the other hand, despite all the strict rules of organic farming, inevitable factors such as atmospheric conditions, soil properties, climatic conditions, continuation of permanent pollutants for years may cause the residues in organic vegetables and cereals thus indirectly (with food intake) results in negative factors/residues in animal products [9].

2.1. Bacterial contaminants in organic animal products

In organic farming, various factors such as use of animal manure, the prohibition of the usage of certain food additives and antibacterials, keeping animals on pasture for longer duration, preferring slow-growing breeds and small slaughterhouses makes organic products vulnerable to bacterial contamination [2, 10, 11]. Studies on bacterial contamination of organically grown animals and animal products are very limited. In fact, concerning the risk of bacterial contamination among organic products, plant products have priority. In terms of organic animal products, poultry meat seems to be more risky. Salmonella and Campylobacter are the most important foodborne bacterial contaminants [10]. Salmonella can lead to disease in humans through consumption of contaminated beef, pork, poultry meat and eggs or vegetables contaminated with animal faces [12]. Differences are seen between the results obtained from the conventional and organic products in terms of contamination with bacteria. In a study, Salmonella was seen in none of the organic chicken farms (layers and broilers), whereas it is evident in approximately 10% of conventional farms, but Campylobacter was observed in all organic broiler farms [13]. Cui et al. [10] analyzed organic and conventional eggs collected from Maryland (USA) retail stores for Campylobacter and Salmonella, and detected Campylobacter in most of the organic (76%) and conventional (74%) chickens and Salmonella was seen in 61 and 44% of organic and conventional chicken, respectively. In the United Kingdom, Campylobacter was found in 80% of organically grown chicken. In a study conducted in Germany, it had been reported that organic chicken meat was contaminated with extended-spectrum beta-lactamase (ESBL) as much as conventional poultry meat [14, 15]. In organic or free-range hen breeding contamination of eggs with the faces and thus the risk of bacterial contamination of eggs is higher than the conventional cage breeding [16]. Antibiotic resistance of the bacteria isolated from organic and conventional chicken and also eggs derived from them differ. In a study, no difference was determined between organic and conventionally grown chickens regarding sensitivity of Campylobacter isolates to antibiotics [15]. In another study investigating antibiotic resistance against Gram-negative bacterial isolates, the resistance in isolates obtained from organically reared chicken is lower because of the limited use of antibiotics in organic farming [17]. Isolates obtained from Campylobacter and Salmonella positive organic chicken eggs were found sensitive to antimicrobial agents, whereas isolates derived from conventional chicken eggs were resistant to five or more antibacterial agents [10]. Similarly, in the Netherlands, antibiotic resistance was lower in microorganisms (except Campylobacter) isolated from faeces samples of organic broilers [13].

It was observed that Salmonella contamination status varies in organic fattening pig farms depending on the breeding experience of the farms [12].

Organically grown animals have a lower risk of bovine spongiform encephalopathy (BSE, mad cow disease) just because they are fed with organic feed [7]. In cattle breeding, there is no basis (evidence) associated with organic production systems in terms of *Escherichia coli* (0157: H7) epidemics. In fact, a meat product such as undercooked minced meat is considered as responsible for the outbreaks due to this microorganism [18]. In a study monitoring the tetracycline residues (tetA and tetB) and tetracycline resistant bacteria in organic meat and vegetable-based baby foods, tet genes have been found in all organic products, particularly higher tetA have been detected in those from poultry origin, which indicates that organic foods are not better than conventional ones [19].

The bacterial count in raw milk is considered as an indicator of hygienic management of the farm. According to the European Union (EU) Council Directive (EC 92/46/EEC) for the production of heat-treated drinking milk, plate count (30°C) for per ml of milk should be ≤100,000, somatic cell count-SSC for per ml of milk should be ≤400,000 in cows' milk and plate count (30°C) for per ml of milk should be ≤1,500,000 in goat's and sheep's milk [20]. In a comparative study, total mesophilic bacteria count-TMBC ($\times10^3$ CFU/mL) and coliform bacteria count-CBC ($\times10^1$ CFU/ml) content of organic milk samples (for mesophilic n = 218; for coliform n = 101) were higher than conventional milk (for mesophilic n = 1168; for coliform n = 473) [21]. In one of the two different studies conducted in USA, no difference was present between organic and conventional (sum of grazing and not grazing) milk regarding SSC [22], and in the other study, very little difference was determined in terms of SSC and standard plate count [23]. Although no difference was found between organic and conventional milk samples concerning the diversity of spore forming aerobic bacteria, bacteria isolated from milk obtained from conventional farms were found to be more resistant to heat, and *B. cereus* organisms were abundant in organic milk, whereas *Ureibacillus thermosphaericus* were abundant in conventional milk. It has been suggested that this situation may be related to dietary strategy in the farm [24], and restricted silage use in organic ruminant breeding may reduce the bacterial contamination (*Listeria monocytogenes, E. coli* O157s) [24, 25].

2.2. Fungal contaminants and mycotoxins in organic animal products

Mycotoxins are toxic molecules, which are synthesized by molds growing on plants. These highly toxic and heat-resistant toxins are transferred to animals with plants, and to humans with animal products through the food chain. Among the mycotoxins, particularly aflatoxin (AFL), ochratoxin (OTA), fumonisins, deoxynivalenol (DON), patulin and zearalenone are the most important mycotoxins for public health. Mycotoxin contamination in animal products is lower than in those from plant origin. Studies comparing the organic and conventional animal products concerning mycotoxin contamination is limited [25].

In Latvia, mold strains belonging to 15 genera were identified in the raw milk samples collected from organic farms between December 2011 and November 2012. Among these strains, the most common ones were *Absidia, Aspergillus, Apophysomyces, Mucor, Penicillium* and *Rhizopus* spp. [26]. In a study of Ghidini et al. [6], Aflatoxin M1 levels in organic (Mean 35 ng/L; Range <5–93 ng/L) was found to be higher than conventional (Mean 21 ng/L; Range <5–66 ng/L) milk samples. The Aflatoxin M1 levels in 49% of the organic and in 10% of conventional milk samples were higher than the legal limit of 50 ng/L, which was set by EU Regulation 466/2001. However, in general, the samples were accepted as safe. In a study analyzing the organic and conventional milk samples for mycotoxins, OTA was detected in 6 out of 40 (11–58 ng/L) conventional milk samples and 5 out of 47 (15–28 ng/L) organic milk samples. OTA was not found in any of 20 baby food. The levels found in milk were higher than 5 ng/kg/day, which is the value for tolerable daily intake-TDI. It has been reported that consumption of such milk would be harmful for children [27]. In Greece, aflatoxin M1 (AFM1) (range 5–10 ng/L) was detected by ELISA in 196 different types (conventional, organic and children's milk) of milk samples collected from the market between November 2009 and June 2010. However, the AFM1 level determined in only two of the samples were higher than the maximal limit set by EU [28]. In a study conducted in Italy, feed and serum of conventional and organic layers and broilers were analyzed, and ochratoxin A (OTA) was found in all of the feed samples (100%). But not above limits set by the EU. OTA rates were high especially in the sera of laying hens on both organic (73%) and conventional (52%) systems, but there was no statistical difference between the laying hens vs broiler group [29].

An OTA contamination (mean 0.05 μg/kg) in organic pork (4/7) was determined by a study conducted in Denmark between 1993 and 1994 [30].

2.3. Metal residues in organic animal products

Although, mineral supplementation in organic animal husbandry is not a routine practice, mineral supplements can be applicable. The diet of the animals in organic farming must be 100% organic [31]. Since organic animals depend on the mineral content in the soil, unlike the expectation, mineral deficiencies can occur in animals. This condition usually results in lower essential elements levels in organic animal products compare to conventional animal products. A study conducted in Spain investigating the levels of essential elements such as Cobalt (Co), Chromium (Cr), Copper (Cu), Ferrous (Fe^{2+}), Iodine (I), Manganese (Mn), Molybdenum (Mo), Nickel (Ni), Selenium (Se) and Zinc (Zn) and toxic elements such as Arsenic (As), Cadmium (Cd), Mercury (Hg) and Lead (Pb) in organic and conventional milk samples has revealed that levels of essential elements in organic milk is lower than conventional milk and toxic element concentrations are lower in both without any significant difference [32]. The analysis of pork obtained from slaughterhouses (n: 20) has shown that As, Pb and Hg (excluding one sample, 0.008 mg/kg) are below the detection limit (0.1, 0.05 and 0.005 mg/kg, respectively). In the same samples, Cd levels were between 0.005 and 0.38 mg/kg (median: 0.11 mg/kg), which were lower than the limits set by the EU (1 mg/kg) [13].

Heavy metals are persistent pollutants like organic chlorine and polychlorinated biphenyl (PCB) pesticides. Since heavy metals still exist in production processes for different purposes, they can be found in high levels in various environmental samples especially in pastures close to industrial areas [6]. Heavy metals enter the body through inhalation of their dust, drinking of the contaminated water or ingestion of the products grown in the contaminated regions (food chain) [33].

Some researchers have shown that levels of harmful elements such as Pb, As, Cd and Ni in organic products are not lower than those found in conventional products [6, 34, 35]. In a study evaluating a total of 156 organic and conventional milk and meat products (78 samples in each group), the mean Pb levels were detected as 1.85 and 1.68 μg/L and the mean Cd levels were detected as 0.09 and 0.16 μg/L in organic and conventional milk samples, respectively. In meat samples, the means of 5.91 and 14.81 μg/L Pb and the means of 0.49 and 1.31 μg/L Cd were detected in organic and conventional samples, respectively. Pb levels in organic and conventional milk samples were not higher than the 20 μg/L, which was set by EU Regulation 466/2001. There is no maximum residue limit (MRL) value for Cd-concerning milk. In the case of meat samples, Pb and Cd levels were lower than 100 and 50 μg/kg, which were set by EU Regulation 466/2001, respectively [6].

In Poland, milk and hair samples obtained from Holstein cow on organic farms were analyzed for Aluminum (Al), As, Barium (Ba), Cd, Cr, Cu, Fe, Hg and Pb, and the mean values of these elements in milk samples were 63.64, 12.27, 26.36, 1.130, 15.76, 157.6, 785.7, 0.396 and 6.210 μg/kg, and the mean values in hairs were 14224, 34.82, 298.7, 2.700, 75.76, 2263, 15925, 82.78 and 32.67 μg/kg, respectively [36].

In Turkey, in a study conducted on milk and milk products offered to consumption between March 2010 and February 2011, samples of conventional and organic products were collected at three monthly intervals and analyzed by Graphite Furnace AAS for Al, As, Cd and phosphorus (P), and the levels of these elements were found lower than limit of detection-LOD values, which were 0.02, 0.001, 0.001 and 0.02 μg/L for Al, As, Cd and Pb, respectively. Mean Pb levels were found as 0.001 ppm in organic milk (n:3) while 0.008 ppm in organic cheeses (n:7). There is not a maximal limit set by Turkish authorities for organic products, therefore, when 0.02 ppm, which was set as maximal acceptable value for the milk according to Turkish Food Codex "Communiqué on Determination of Maximum Level of Certain Contaminants" (Communiqué No: 2002/63) in foodstuffs, was taken as the basis, the Pb levels determined in one organic cheese and in one organic butter were above the maximum acceptable level [37, 38].

In a study performed in Turkey (Aegean Region) for determining the mineral content of the organic and conventional chicken eggs, compare to conventional chicken eggs, P and Zn levels in the edible portion of organic chicken eggs were lower, whereas Mg was higher in shell, and there was no difference between organic and conventional eggs concerning calcium (Ca), Fe and Cu contents [39]. In Turkey, 0.020, 0.055 and 0.020 mg/L of Cd, Pb and Cu, respectively, were found below the LOD in all of the organically and conventionally produced flower honey and eggs by analysis, whereas Fe concentrations were found at higher levels in organic compared to conventional products [40].

In Greece, in a controlled study, Cu, Vanadium (V), Cr, Ni, As and Cd contents were determined in conventional, organic and free-range (in the courtyard) chicken eggs, and mean values for these elements were determined as 1357, 12.5, 66.2, 63.3, 13.9, 1.4 ng/g in conventional, as 1233, 13.2, 82.9, 58.4, 12.5, 1.6 ng/g in organic and as 1282, 12.6, 90.5, 59.2, 15.4 and 1.5 ng/g in free-range chicken eggs, respectively. The values were lower in white than those in brown eggs [41]. In Egypt, in the analysis of organic eggs for Cd, Pb and Al showed that Cd and Pb were present in 34 and 40% of the organic eggs, respectively. The Cd and Pb contents of the eggs were above the maximum permissible levels. It was emphasized that although, when calculated according to target hazard quotients (THQ) organic eggs appear to have a low health risk, they are not safer than conventional ones [42].

Analysis of Cd levels in liver, kidney and fecal samples as well as feed, soil and water samples collected from a pig farm in which organic (outdoor) and conventional (indoor) breeding systems implemented together showed that Cd levels in organically and conventionally produced feedstuffs were 39.9 and 51.8 μg/kg, respectively. Cadmium content in 38% of the feed given to conventionally reared animals was found to be associated with the Cd content of beet fibers, which was included in to diet at a rate of 5%. No difference was determined between liver samples collected from the animals on organic and conventional feeding systems concerning Cd levels (15.4 ± 3.0 μg/kg). Despite the low amount of Cd in feed, more Cd was found in kidney of pigs fed with organic feed. In addition, Cd levels were higher in the feces of organic pigs, which were attributed to environmental exposures such as soil [43].

2.4. Pesticide residues in organic animal products

Organic products contain more phytochemicals, which are protective against pests, therefore, use of pesticides is not required, thus the risk of pesticide residues in organic products is low

[34]. However, from time to time, pesticides such as DDT and its degradation products, DDE may be found in foods such as organic-grown grain, grain products (biscuits, bread, etc.), meat and dairy products. Despite the use of pesticides in organic farming is not allowed, the reason for the existence of these substances in organic farming is attributed to the ability of them to remain in environment for a long time without disintegration [44]. Pesticides can be encountered in most of the animal products (meat, milk and fish) depending on bioaccumulation. Dioxin and dioxin-like compounds (polychlorinated dibenzo-p-dioxins-PCDD, polychlorinated dibenzofurans-PCDF, and polychlorinated biphenyls-PCBs), which are a general problem of the places in where industrialization is intense or intensely populated create similar problems for organic or conventional farming [45]. It has been stated that 4% of dioxin received by people per day comes from the eggs [46]. It has also been claimed that more dioxin was determined in eggs obtained from free-range hens compared to those obtained from hens grown indoors [45]. In a study conducted on honey for the evaluation of pesticides, it has been emphasized that there is no significant difference between organically and conventionally produced honeys [47]. In northern Italy, in the analysis of conventional and organic animal meat and milk samples for organochlorine pesticides and PCBs, pesticides and PCBs have been found below legal limits in both organic and conventional samples [6]. In another study conducted in Italy, the residues of persistent pollutants and pesticides were determined by GC-MS/MS analysis in most of the 59 organic honey samples. However, levels were below the MRL. This result was attributed to geographical conditions [48]. According to the findings of the United States Department of Agriculture (USDA) pesticide data program (PDP), the market place surveillance program of the California Dept. of Pesticide Regulation (CDPR) and a Consumers Union private residue-testing program, conventional/organic pesticide residue ratios have been found as 3.2, 4.8 and 2.9, respectively. These results seem to relieve the 70% of people who prefer the consumption of organic products to avoid from pesticides [49].

2.5. Veterinary drug residues in organic animal products

Outdoor rearing of animals in organic livestock production system may increase the risk of animals to contact with environmental pathogens that cause foot diseases (especially in pigs) as well as infectious diseases and helminthiasis. Lack of use of the curative and preventive conventional medicines (antibiotics) in organic farming leads to concerns about the treatment of the diseases. Mastitis is one of the most common diseases seen in dairy animals. Mastitis incidence is reported to be higher in organic production than in conventional production in England, Germany, to a lesser extent in Norway, Sweden and Denmark. However, it has been indicated that the difference between conventional and organic farming is decreased with the increasing awareness of animal production [4]. Since the use of veterinary drugs has not been allowed in organic livestock production, screening organic animal products for veterinary drug residues is at negligible levels. However, in a study conducted by Ghidini et al. [6], the antibiotic residues have been found at low levels, approximately 0.3%, in milk. In a study conducted before 1997, no difference was reported between conventional and organic honey in terms of veterinary drug residues [47]. In the analysis of kidney and meat samples of organic pigs (n = 20) taken from slaughterhouses, solely one sample showed a slight bacterial inhibitory effect against macrolide antibiotics. All of the organic and conventional eggs were found negative by the analyses for toltrazuril aminoglycosides, sulfonamides, beta-lactam antibiotics, tetracyclines, quinolones and colistin residues [13].

3. Contaminants in conventional foods

Developments in medicine, industry and agriculture have caused the world's population to increase and as a result of the need to feed the increasing population and urbanization, it became a necessity to produce more in agriculture and industry. This necessity caused widely use of different chemicals (pesticides, heavy metals, veterinary medicines, etc.) in various areas of production and this caused high amount of disposals of wastes in an uncontrolled manner to the environment, which results pollution. In parallel with the increase in pollution, the contaminants in food resources caused significant health problems in humans as a result of food chain [50].

3.1. Metal residues in conventional animal products

Heavy metals are the elements with an atomic weight between 63,546 and 200,590 and with a specific gravity more than 4.0 [51]. Metals are dispersed in the nature through geological and biological cycles [52] and then can penetrate to the food chain by contaminating the cereals from the environment, the animals and animal products from contaminated cereals and herbs, and fish from the polluted waters [52, 53].

Metals have harmful effects on most of our organs due to their elementary structure and their affination with organic ligands through biological cycles. Since metals are strongly bound to tissues, they are disposed very slowly and accumulated in the body. Samples of blood, urine and hair are usually used as indicators in evaluating the level of exposure to metals [52]. Although soil is the primary source of toxic metals in edible plants, the level of contamination increases more with metal wastes, consumption heavy metal wastes, leaded fuels and paints, fertilization of soil, animal fertilizers, sewage wastes, pesticides, irrigation with waste water, wastes of coal burning, spillage of petrochemicals, atmospheric accumulation, volcanic activities, etc. [54, 55]. It was revealed in the study of International Atomic Energy Agency (IAEA), which was conducted on various food samples taken from 12 countries, that Pb, Cd, Hg and As are important in terms of health and contamination risk, whereas antimony (Sb), Fe, Cu and Zn are less important [56]. International Agency for Research on Cancer (IARC) has specified Cd and Cd components as Group I carcinogen for human health (they induce lung tumors) [57]. Heavy metals, such as As, Cd, Hg, Cu, Pb, etc. that contaminate water through any means can accumulate within fish and then cause health problems in humans [58].

Maximum limits of Cd, Pb and Hg in some animal products are given in European Commission (EC) and Turkish Food Codex (TFC) [56]. There is also information about daily consumption amounts of metals that humans can take. Daily consumption amount of Sb is specified as 0.25–1.25 mg for children in the USA. The USA has determined that Al consumption should not exceed 12–14 mg/day for young and adult men and 9 mg/day for young and adult women [52]. Although, Zn is an essential element for human body, according to animal experiments, high doses of Zn is toxic and carcinogenic [53, 59]. The amount that can be taken with food

http://dx.doi.org/10.5772/67064

is set as 0.23 mg/kg/day by the FDA [59]. Contaminated seafood with industrial wastes may contain high level of Zn, and entry of these products into the food chain can pose a danger to human health. Food and Agriculture Organization (FAO) and World Health Organization (WHO) and has determined that maximum amount of daily allowable consumption of as should be 2 μg/kg of body weight [60].

Most of the foods other than fish contain <0.25 mg/g As, but many fish species contain As between 1 and 10 mg/g. However, the amount of As found in marine crustaceans and deep sea fish was found as 100 mg/g or more [51]. Although the amount of As consumed is 10–200 μg/day, this amount can reach to several thousand μg/day in those that consume fish a lot in their diet [58]. The accumulated amount of As is 3–10 ppm in oyster, 42–174 ppm in mussels and 42–174 ppm in shrimps. Thus, most of the As taken with food by human is originated from sea foods [51]. However, As poisoning due to consumption of animal products is also possible. It was seen in early summer of 1955 that the babies younger than 12 months in western Japan had symptoms of anorexia, skin pigmentation, diarrhea, vomiting and distention and more than 100 babies showing these symptoms died and then it was found that the case was caused by consumption of powdered milk (popular and brand), which contains approximately 21–34 μg As per gram and As was found in the babies that consumed this powdered milk. It was also found that the origin of As was disodium phosphate, which was added to cow milks as a stabilizer [51, 61].

Cadmium, one of 25 substances that have a certain potential of danger against human health, cannot be disposed from and is accumulated within the body [57]. Foods usually contain Cd less than 0.05 ppm. However, WHO announced that the highest level of Cd was found in crustaceans as well as the kidneys of various animals, such as cow, chicken, pig, sheep and turkey as a result of analyses. Daily tolerable amount of consumption of Cd is 1 μg/kg of body weight [52]. The US Environmental Protection Agency (US EPA) stated that Hg and Hg components, which cause kidney cancer in experimental animals, may also cause cancer in humans [53]. The amount of Hg ranges between <1 and 50 μg in many food and beverage. However, the most important source of Hg in diet is the fisheries, caught from contaminated waters. Since crustaceans, such as mussel and oyster, feed by filtering water, they accumulate Hg components in their bodies. Mercury exists in bigger fishes in higher concentrations, compared to smaller ones. According to a research conducted by FDA, the amount of Hg in big tuna fishes was 0.25 ppm, whereas it was found as 0.13 ppm in average in smaller tuna fishes. The type of Hg that is found most in sea foods (>90%) is methyl mercury. FDA determined maximum allowable level of Hg in fish and crustaceans as 0.5 ppm [51]. Methyl mercury poisoning or Minamata disease, seen in Japan in 1954, is the most important example of Hg poisoning due to animal products. This disease was caused by consumption of fishes, living in water that was heavily contaminated with industrial wastewater. Similarly, serious muscle and neurological dysfunctions were seen in humans living in the city of Nigata and close to Minamata Bay in 1970 and 50 of 120 hospitalized persons died [51, 62].

3.2. Pesticide residues in conventional animal products

Pesticides are chemicals, most of which are highly toxic and are used against pests. These substances are toxically effective not only against pests but also other living organisms. Pesticides cause behavior disorders, immunosuppression, allergic reactions, estrogenic, teratogenec, mutagenic and genotoxic effects on living organisms. The duration of stay of pesticides in the natural environment, depend on their chemical structure. Pesticides, such as chlorinated hydrocarbons are resistant against biological degradation and they can stay in soil for years and penetrate to the food chain through various means. These fat soluble pesticides can be accumulated in the fat tissue of humans and animals as well as in their livers, kidneys and neural systems. Residues in the body of lactating animals can easily penetrate to the animal's milk [51, 52]. Contamination of animal products, such as meat and milk, with permanent pesticide residues is a frequently encountered problem. In a study conducted in Jordan, in which eggs as well as meats of chicken, sheep and cow were scanned for OCP residues, it was found that 28% of eggs, 20% of chicken meat, and 49% of red meat were contaminated with OCP [63].

Chlorinated compounds, such as PCBs, aldrin, DDT, DDD, DDE, BHC, heptachlor, etc., which enter the body of fishes through various means, can accumulate within the fishes and cause health problems in humans that consume these fishes. There is a linear relationship between accumulation of chlorinated compounds, such as PCBs, within fishes and their fat contents. The experiments showed that half-lives of PCBs in fishes are quite long. Despite the fact that utilization of PCBs was banned, they were still found in fish samples, analyzed in Ontario, Canada in 1992–1993, and in fish samples collected from 15 different countries in 1994–1995 [58].

3.3. Bacterial contaminants in conventional animal products

Milk is considered sterile (free from microorganism) because of its compounds and chemical properties. But milk is a suitable medium for most microorganisms. In general, it is not expected that milk has microorganisms and toxins unless there is a systemic or local infection. But clinical and subclinical mastitis, which are associated with local or systemic infections are common problems for animals [64, 65]. The milk flora of dairy animals consists of lactic acid bacteria (LAB; *Lactococcus, Lactobacillus, Leuconostoc, Streptococcus, Enterococcus* spp.) [64]. *Staphylococcus aureus,* which produces toxins like Staphylococcal enterotoxins (SEs), SE-like toxins (SEl) and toxic shock syndrome toxin (TSST-1) and is primarily responsible for foodborne poisonings, mostly exists in milks of animals with mastitis [65]. According to State Agencies to Centers for Disease Control and Prevention and from the Center for Science in the Public Interest Database, product-based numbers of *Campylobacter*-based cases caused by consumption of raw milk, pasteurized milk and cheese that's produced from raw milk between 2000 and 2006 were recorded, respectively, as 33, 1 and 3; numbers of *E. coli*-based cases were recorded, respectively, as 6, 0 and 1; numbers of *Salmonella*-based cases were recorded, respectively, as 1, 3 and 3 [66]. In a study that was made in Ankara (Turkey) with milk collected from street mostly found *S. aureus* > *E. coli* > *Klebsiella* > *Serratia* > *Proteus* [67]. In a study conducted in Czech Republic, total amount of mesophilic bacteria-TMBC ($\times 10^3$ CFU/ml) in conventional milk was found as 19 ± 16 (as Mean ± SD; n:1168) and amount of coliform bacteria-CBC ($\times 10^1$ CFU/ml) was found as 48 ± 36 (as Mean ± SD; n: 473) [21]. In low input farms in Brazil, bulk milk bacteria count (BMBC) was found higher in winter 2174 ± 958.4 (Mean ± SEM) according to other seasons. But in same season bulk tank somatic cell count (BTSCC × 1000 cells/ml) was found as 469 ± 113.4 (Mean ± SEM) [68]. In a study with raw milk in winter and summer In Slovenia, total bacteria count was found higher than 100,000 cfu/ml [69].

3.4. Fungal contaminants and mycotoxins in conventional animal products

Mycotoxins are very toxic compounds that are produced by fungi and yeast [70]. Diseases due to the consumption of contaminated food with mycotoxins and molds are known worldwide. Grain and milk products are the most sensitive ones to contamination with mycotoxins among foods [71]. In mycotoxicosis cases, consumption of animal products (milk and dairy products, meat and meat products, egg, liver, kidney) has a major role as well as consumption of grain and grain products. Mycotoxins cause respiratory and neurological disorders, cancer, nephrotoxicity and hepatoxicity. Diseases such as Alzheimer's, multiple scleroses, etc. are considered to be related to mycotoxicosis. In pregnant women, mycotoxins that are taken with contaminated products can affect baby through placenta. Especially, infant and children are very sensitive to mycotoxins [72]. As a result of research in infant foods (rice flour, grain flour and milk powder) *Aspergillus* spp. (5%), *Penicillum* spp. (13%), *Mucor* spp. (5%) and unidentified species were isolated [71].

First mycotoxin (aflatoxin M1) contamination in dairy products was recorded in 1960s. Aflatoxin M1 (AFM1) is a metabolite of aflatoxin B1 (AFB1) and it forms in liver. 0.3–6.2% of AFB1 in animal feeds is metabolized, biotransformed, and secreted in milk in the form of AFM1. Mycotoxins such as OTA, zearalenone (ZEN), T-2 toxin and DON were also detected in milk. But these are not taken into account in importance as much as AFB1. One of the main reasons of DON and ZEN contamination is silage that is added into animal food [70]. Contamination with fungi and mycotoxin formation are not necessarily related to each other. Even when fungi contamination and variety is high mycotoxin can form less. According to a research in infant foods aflatoxin was detected only in 2.4% (19–70 μg/kg) of specimens despite of high fungus contamination [71]. In the European Union and some other countries accepted limits of AFM1 varies for raw milk is between 0.05 and 10 μg/kg, for dairy products is between 0.02 and 10 μg/kg [70].

In a study where fungal contamination variety's being analyzed of cow, goat and sheep milk, turned out that cow milk samples the highest diversity, and it was recorded that identified species were belonged to *Aspergillus, Chrysosporium, Cladosporium, Engyodontium, Fusarium, Penicillium* and *Torrubiella* genera [73]. There are less yeast and mold in raw milk than LAB [64]. In a study conducted in Slovenia, 95.0% of raw milk that was collected during winter and summer contains yeast and 63.3% contains molds. Isolated mold strains were identified as *Geotrichum* (51.5%), *Aspergillus* (33.8%), *Mucor* (5.9%), *Fusarium* (2.9%) and *Penicillium* (2.9%) genuses [69].

Poultry meat can also be contaminated by mycotoxins. A study showed that most common mold genuses are *Aspergillus* (58%) and *Penicillium*. Also, many other fungus genuses had been found with low incidence [74].

3.5. Contaminants in conventional animal foods from packaging material

Packaging is an indispensable part of the food production process. Today lots of plastics are being used as packaging material. Also, antioxidants, stabilizers, lubricants, antistatic and antiblocker materials can be used to increase the performance of package material. Additives, monomers, oligomers and contaminants can get transferred to food from packaging material. There are concerns about plasticizers (phthalates), thermal stabilizers, slip additives, light stabilizers, antioxidants, melamine, styrene, vinyl chloride, bisphenol A diglycidyl ether, isocyanate, caprolactam, polyethylene terephthalate oligomer, decomposition products, benzene and other volatiles, environmental contaminants, processing agents and other contaminants getting transferred to food [75, 76]. Studies on contamination in milk products related to this issue are limited [76].

Especially heavy metal pollution can occur in canned milk products and this is related to storage temperature and duration [77]. Also heavy metal pollution can occur during packaging process. As a result of a study, high amount of Pb was detected on bread packages [78]. In another study, high amount of Pb was detected on candy packages, which children consumed often, and this result was backed up by FDA [79, 80].

Because of that Cd got high dissolution in organic acids, human food chain's Cd pollution is very common. Studies showed that Cd, which is used for making food packages, can get transferred to high-acidic foods by getting dissolved. Wrapping foods with antimony foil, keeping in antimony containers and cooking in them causes foods get contaminated with high amount of Sb [51, 52]. Zinc can get transferred through galvanized containers to humans [56].

3.6. Veterinary drug residues in conventional animal products

Nowadays, various veterinary drugs and food additives are being used as therapeutic and prophylactic in animals. Foods of animal origin that contains drug residue consumed by human can cause allergic reactions, drug resistant microorganisms, toxicities in organs and tissue, hormonal disorders, teratogenic effects, etc. Animal originated milk and dairy products can contain veterinary drug residues as contaminants such as antimicrobials (like antibiotics), hormones, anthelmintics and pesticides. Beta-lactams, tetracyclines, aminoglycosides, macrolides and sulfonamides are the most commonly used antibiotics [81]. The result of a study made by USDA showed 5.3% of 529 carcasses have antibiotic residue. In these tests, chlortetracyclines, oxytetracyline, tetracycline, streptomycin, neomycin and erythromycin antibiotics had been detected [52]. In a study conducted in Croatia, 1259 raw milk samples were analyzed for antibiotic residue (chloramphenicol, penicillins, cephalosporins, tetracyclines, sulfonamides, beta-lactams, quinolones, aminoglycosides and macrolides) and 37 positive samples were found, but because of low levels it was stated that this would not cause any health problems [82]. The usage of chloramphenicol, which causes bone marrow suppression and aplastic anemia, is prohibited for animals. In Brazil where its usage prohibited in 1998, study made with ELISA showed 28.6% 84 raw milk samples were positive for chloramphenicol [83]. In Egypt, after antibiotic analyzes on broiler fillets, which were collected from markets, it turned out there were problems especially about detecting withdrawal times of oxytetracyline residues [84].

Steroid hormone can be in milk. Food production processes do not have any effect on milk and dairy products. Testosterone was detected in fresh cheese (0.1–0.5 mg/kg). Benzimidazole anthelmintics are being used commonly on animals thus benzimidazole anthelmintics and their metabolites (albendazole sulfoxide, albendazole sulfone, etc.) can be in dairy products [81]. The result of a study conducted in Macedonia analyzes showed only one of 55 bovine meat samples was positive for clenbuterol [85]. For preventing and curing diseases in fishes, veterinary drugs such as antibiotics mainly, anthelmintics and hormones are being used. Sometimes nonprescription or prohibited drugs can be used. In Canada, after analyzes made with sea, fresh water and canned fish, as ng/g level furazolidone metabolite 3-amino-2-oxazolidinone (AOZ), enrofloxacin, leucomalachite green, oxolinic acid and chloramphenicol residues were detected. In 28 eel samples, which were collected from markets in Tokyo, 0.07 ppm oxolinic acid was detected. Again in Tokyo, in flounder sample, which was collected from markets, 360 μg/kg oxytetracyline was detected on the skin [86].

4. The effect of various cooking and freezing processes on contaminants in animal foods

In the case of therapeutic drugs, before using the product, implementation of withdrawal time for the drug residues has been made mandatory. The obligation of drug applications to sick animals requires the disposal of the products containing residues of during this period, which means economic losses. Withdrawal time of drug residues in animal products is usually determined on unprocessed products. However, most of the animal products are consumed after certain treatments (such as cooking or storing in cooler at a certain time). Such processes may affect the drug residues in the products. Some previous studies have shown that processes applied to the product containing residue may result in changes in the level (quantity) of drug residues [87–90]. This suggests that, in inevitable conditions, the product containing residue is subjected to conditional consumption. Most of the researches on the subject are related to conventional animal products. The obtained results may vary depending on various factors such as quality of the animal products, the sample site on the same animal, the kind and duration of the applied processes. Studies have shown that tetracycline residues were decreased by 35–94% in muscle (cattle and sheep) and liver (cattle) through cooking (microwave, boiling, roasting, grilling and frying). Residues of penicillin (penicillin G-benzylpenicillin and cloxacillin) in milk have been reported to be decreased by the boiling and yogurt production (fermentation). On the other hand, since penicillinase released by microorganisms found in raw milk is deteriorated in the milk produced by UHT, benzylpenicillin is more stable (not disintegrated) in milk produced with this technique. Cooking cannot reduce the residues of oxolinic acid, flumequine, enrofloxacin and ciprofloxacin, which are belonging to Quinolone group, in fish. However, such residues can be removed by discarding the meat broth containing the residues, which are transferred into boiling water through boiling [87]. A similar situation has been observed in broilers concerning some drugs belonging to sulfonamides (sulfadiazine) and quinolone (danofloxacin) groups [88, 89]. Cooking decreases sulfamethazine residues in tissues (muscles and liver) of broiler at different rates. The most significant decrease occurs in boiling because during the boiling process drug in the tissue passes to water. Similarly, cooking (boiling and grilling, equally effective) may also be effective on sulfachloropyridazine-trimethoprim combinations in broiler tissues (muscle and liver) but these drugs cannot be transferred into boiled water in contrast to sulfamethazine [89]. Concentration of levamisole residues in broiler tissues (muscle, liver) can be diminished by different cooking processes (through disintegration and passing to water), whereas the effectiveness of deep freezing is time-dependent and the most losses occur on day 30th [90].

Especially washing as well as applications such as chlorine, chlorine dioxide, hydrogen peroxide, ozone, acetic acid, peracetic acid, hydroxy, iprodione can significantly reduce the pesticide residues in foods. Processes such as pasteurization, boiling, steaming and canning can reduce the levels of pesticide residues depending on the treatment type and time as in veterinary drug residues. In contrast, the implementation of food preservation techniques such as drying or dehydration increases the concentration of pesticides (due to a reduction in weight of product resulting from drying) [91].

Except the studies investigating the effects of processing on pesticide residues mostly in vegetables and cereals processing have diverse effects on pesticide residues in animal products such as milk (pasteurization) dairy products (cheese and yoghurt production) and eggs (boiling and scrambling). When reduction in pesticide residues in dairy products were compared, the reduction in foods made of sheep and goat's milk may be 50% less than in those made of cow's milk. Hexachlorocyclohexane (HCH) residues show a gradual decline by yoghurt production and by keeping at refrigerator [91]. Sausage making can lead to a significant reduction in organochlorine (hexachlorobenzene-HCB, α-, β-, γ-hexachlorocyclohexane-HCH and p,p'-DDE) pesticide residues [92].

Accumulation of organochlorine insecticides in fish is 10–10,000 fold higher than water [52]. Boling process is very effective in reducing DDT and heptachlor concentrations in dried fish (Bombay duck-loittya, ribbon fish-chhuri, shrimp-chingri, Chinese pomfret-rupchanda and Indian salmon-lakhua) [93]. It has been reported that frying process is effective in reducing α-HCH, β-HCH, γ-HCH, heptachlor, aldrin, heptachlor epoxide isomer B, pp'-DDE, endrin and pp'-DDT residues in commonly consumed fish (*Clupea harengus* L., *Salmo salar* L., *Cyprinus carpio* L., *Salmotrutta m. fario* L., *Platichthys flesus* L. and *Gadus morhua* L.) in Poland, and the most pronounced reduction is observed in β-HCH residues [94].

5. Legal regulation for foods of animal origin

Maximum residue limit (MRL) is defined as the highest concentration of a chemical residue that is legally permitted or accepted in a food, and acceptable daily intake (ADI) is defined as the amount of a residue that can be ingested on a daily basis over a lifetime without health risk [52]. National/international information concerning the maximum level of contaminants allowed in conventional product is available. Maximum levels for contaminants in conventional food of animal origin were determined by the EU. European Food Safety Authority (EFSA) makes risk assessment for pesticides and European Commission determines appropriate MRLs [95]. Food Additives FAO/WHO Joint Expert Committee (JECFA) determines the tolerable weekly intake levels of heavy metals in order to prevent heavy metal contamination in foods whereas EFSA and the Codex Alimentarius Commission (CAC) offer proposals for the exposure and tolerance limits of the heavy metals [50]. The EU directive No. 1881/2006 setting maximum levels for certain contaminants in foodstuffs entered into force following its publication in 2006. The MRLs in foodstuffs for nitrates, mycotoxins, metals, 3-monochloropropenes-1,2-diol (3-MCPD), dioxins and PCBs, polycyclic aromatic hydrocarbons (PAH) are specified in the relevant directive. The veterinary drug residue limits (MRLs) for a variety of foods including animal origin are determined by Committee for Medicinal Products for Veterinary Use (CVMP) of the European Medicines Agency (EMA) [95]. The first directive that concern to protect consumers from harmful substances coming from packaging materials was published by Commission of the European Communities (CEC) in 1976. Analysis methods for the official control of the vinyl chloride monomer levels in food packaging materials were identified in 1980. According to the regulations made by the EU, countries can make their own private arrangements at the national level [75].

The beginning of legal regulations on organic farming dates back to the 1970s. Studies conducted, independently, on organic farming in different countries became organized under a roof with the establishment of International Federation of Organic Agriculture Movement (IFOAM) that was headquartered in Germany in 1972. IFOAM is the first organization that defines the rules for ecological production worldwide. The rules, initially developed as the series of Basic Principles were modified as IFOAM Basic Standards, adopted by the General Assembly and entered into force in 1998 [96].

The first EU directive relevant to organic products was published on June 24, 1991. This directive, No 2092/91, was established solely for organic vegetable production [97]. In 1999, EU directive on animal production and general standards, "Codex Alimentarius", that was jointly prepared by the FAO and the WHO was published. The Codex Committee on Food Labelling, which was under CAC, lays down the standards pertaining to organically produced and labeled herbal and animal foodstuffs. Moreover, standards deal with plants and plant products, livestock and animal products, sources of animals, the prevention and treatment of animal diseases, such as fertilizer and pest management issues have been implemented [98]. In the following years, directives with different scopes and contents have been prepared and entered into force by the EU [97]. Directives issued by the EU are either accepted as they are by the countries of world or adopted according to their national conditions to create their own regulations.

The presence of any contaminant in organic products is normally not expected due to strict principles of organic farming. However, because some substances are the natural ingredients of the earth, they can be found naturally in organic products like happens in the elements (copper, iron, etc.). The levels of these substances in organic human and animal food (feed and feed ingredients) can vary depending on various factors such as geographical conditions and soil properties. On the other hand, despite the high precision of the organic farming, persistent environmental contaminants resulting from industrial and other activities can be involuntarily reflected in the organic products [99]. Legal regulations regarding the evaluation of organic products for contaminants are considered to be in their early stages. Although this situation varies among countries of the world, the EU seeks to create long-term control programs, especially, on pesticide residues with the issued regulations [100].

6. Conclusion

Food contaminants can cause consumer illness such as allergy, immunosuppression, cancer, teratogenicity, mutagenicity, genotoxicity. Therefore, monitoring of food contaminant is an important issue for the protection of public health. In order to protect public health use of many veterinary drugs for prophylactic purposes is prohibited by most of the countries. However, significant differences can arise among countries concerning the types of prohibited drugs and MRL values. This situation results in problems particularly for imported/exported products. On the other hand, there are still some veterinary drugs that have no MRLs for even conventional animal products. In addition, animal products may include environmental contaminants associated with industrial and agricultural activities. This situation raises concerns about the presence of residues/contaminants in animal products despite strict policy of the legal authorities. Therefore, people, especially in developed countries, tended to consume organic products. However, difficulties in production of organic products thus their high prices result in the consumption of them by only certain populations, which leads to social inequality in society. On the other hand, contamination may arise due to the failure to provide the required standard in organic products. Therefore, the regulative arrangements that are launched by the EU for organic products should be expanded and put into practice at countries basis as in conventional products.

Author details

Emine Baydan[1*], Murat Kanbur[2], Emre Arslanbaş[3], Farah Gönül Aydın[1], Semra Gürbüz[4] and Muhammed Yasin Tekeli[2]

*Address all correspondence to: emine.baydan@veterinary.ankara.edu.tr

1 Department of Pharmacology and Toxicology, Faculty of Veterinary Medicine, University of Ankara, Ankara, Turkey

2 Department of Pharmacology and Toxicology, Faculty of Veterinary Medicine, University of Erciyes, Kayseri, Turkey

3 Department of Pharmacology and Toxicology, Faculty of Veterinary Medicine, University of Cumhuriyet, Sivas, Turkey

4 Higher School of Tourism and Hotel Management, Gastronomy and Culinary Art, University of Mardin Artuklu, Mardin, Turkey

References

[1] Mansour SA, Belal MH, Abou-Arab AAK, Gad MF. Monitoring of pesticides and heavy metals in cucumber fruits produced from different farming systems. Chemosphere. 2009;**75**:601–609. doi:10.1016/j.chemosphere.2009.01.058

[2] Kouba M. Quality of organic animal products. Livestock Production Science. 2003;**80**:33–40. doi:10.1016/S0301-6226(02)00318-4

[3] Krejčová A, Návesník J, Jičínská J, Černohorský T. An elemental analysis of conventionally, organically and self-grown carrots. Food Chemistry. 2016;**192**:242–249. doi:10.1016/j.foodchem.2015.07.008

[4] von Borell E, Sorensen JT. Organic livestock production in Europe: aims, rules and trends with special emphasis on animal health and welfare. Livestock Production Science. 2004;**90**:3–9. doi:10.1016/j.livprodsci.2004.07.003

[5] Magkos F, Arvaniti F, Zampelas A. Putting the safety of organic food into perspective. Nutrition Research Reviews. 2003;**16**:211–221. doi:10.1079/NRR200361

[6] Ghidini S, Zanardi E, Battaglia A, Varisco G, Ferretti E, Campanini G, Chizzolini R. Comparison of contaminant and residue levels in organic and conventional milk and meat products from Northern Italy. Food Additives and Contaminants. 2005;**22**:9–14. doi:10.1080/02652030400027995

[7] Pathak PK, Chander M, Biswas AK. Organic meat: an overview. Asian-Australasian Journal of Animal Sciences. 2003;**16**:1230–1237.

[8] Malmauret L, Parent-Massin D, Hardy JL, Verger P. Contaminants in organic and conventional foodstuffs in France. Food Additives & Contaminants. 2002;**19**:524–532. doi:10.1080/02652030210123878

[9] Pandey J, Pandey U. Accumulation of heavy metals in dietary vegetables and cultivated soil horizon in organic farming system in relation to atmospheric deposition in a seasonally dry tropical region of India. Environmental Monitoring and Assessment. 2009;**148**:61–74. doi:10.1007/s10661-007-0139-8

[10] Cui S, GeB, Zheng J, Meng J. Prevalence and antimicrobial resistance of *Campylobacter spp.* and *Salmonella* serovars in organic chickens from Maryland retail stores. Applied and Environmental Microbiology. 2005;**71**:4108–4111. doi:10.1128/AEM.71.7.4108-4111.2005

[11] Miranda JM, Mondragon A, Vazquez BI, Fente CA, Cepeda A, Franco CM. Influence of farming methods on microbiological contamination and prevalence of resistance to antimicrobial drugs in isolates from beef. Meat Science. 2009;**82**:284–288. doi:10.1016/j.meatsci.2009.01.020

[12] Gómez-Aldapa CA, Torres-Vitela MR, Villarruel-López A, Castro-Rosas J. The Role of Foods in Salmonella Infections. In: Mahmoud BSM, editors. Salmonella-A Dangerous Foodborne Pathogen. InTech, Croatia; 2012. pp. 21–47. doi:10.5772/28316

[13] Hoogenboom LAP, Bokhorst JG, Northolt MD, van de Vijver LPL, Broex NJG, Mevius DJ, Meijs JAC, van der Roest J. Contaminants and microorganisms in Dutch organic food products: a comparison with conventional products. Food Additives and Contaminants. 2008;**25**:1195–1207. doi:10.1080/02652030802014930

[14] Kola A, Kohler C, Pfeifer Y, Schwab F, Kühn K, Schulz K, Balau V, Breitbach K, Bast A, Witte W, Gastmeier P, Steinmetz I. High prevalence of extended-spectrum-β-lactamase producing Enterobacteriaceae in organic and conventional retail chicken meat, Germany. Journal of Antimicrobial Chemotherapy. 2012;**67**:2631–2634. doi:10.1093/jac/dks295

[15] van Loo EJ, Alali W, Ricke SC. Food safety and organic meats. Annual Review of Food Science and Technology. 2012;**3**:203–225. doi:10.1146/annurev-food-022811-101158

[16] Stanley VG, Nelson D, Daley MB. Evaluation of two laying systems (floor vs. cage) on egg production, quality, and safety. Agrotechnol. 2013;**2**:109. doi:10.4172/2168-9881.1000109

[17] Schwaiger K, Schmied EMV, Bauer J. Comparative analysis on antibiotic resistance characteristics of *Listeria spp.* and *Enterococcus spp.* isolated from laying hens and eggs in conventional and organic keeping systems in Bavaria, Germany. Zoonoses and Public Health. 2010;**57**:171–180. doi:10.1111/j.1863-2378.2008.01229.x

[18] Soon JM, Chadd SA, Baines RN. *Escherichia coli* O157:H7 in beef cattle: on farm contamination and pre-slaughter control methods. Animal Health Research Reviews. 2011;**12**:197–211. doi:10.1017/S1466252311000132

[19] Guarddon M, Miranda JM, Vazquez BI, Cepeda A, Franco CM. Assessment of tetracyclines residues and tetracycline resistant bacteria in conventional and organic baby foods. Foods. 2015;**4**:306–317. doi:10.3390/foods4030306

[20] Barth K. Milk quality in organic farming: cows, goats and sheep. In: İzmir Regional Chamber of Veterinary Medicine 1st International Congress on Organic Animal Production and Food Safety; 28 April–1 May 2004; Kuşadası, Türkiye

[21] KouřimsKá L, LegaroVa V, PanoVsKa Z, PaneK J. Quality of cows' milk from organic and conventional farming. Czech Journal of Food Sciences 2014;**4**:398–405.

[22] Cicconi-Hogan KM, Gamroth M, Richert R, Ruegg PL, Stigbauer KE, Schukken YH. Associations of risk factors with somatic cell count in bulk tank milk on organic and conventional dairy farms in the United States. Journal of Dairy Science. 2013;**96**:3689–3702. doi:10.3168/jds.2012-6271

[23] Stiglbauer KE, Cicconi-Hogan KM, Richert R, Schukken YH, Ruegg PL, Gamroth M. Assessment of herd management on organic and conventional dairy farms in the United States.Journal of Dairy Science. 2013;**96**:1290–300. doi:10.3168/jds.2012-5845

[24] Coorevits A, De Jonqhe V, Vandroemme J, Reekmans R, Heyrman J, Messens W, De Vos P, Heyndrickx M. Comparative analysis of the diversity of aerobic spore-forming bacteria in raw milk from organic and conventional dairy farms. Systematic and Applied Microbiology. 2008;**31**:126–40. doi:10.1016/j.syapm.2008.03.002

[25] Lairon D. Nutritional quality and safety of organic food. A review. Agronomy for Sustainable Development. 2010;**30**:33–41. doi:10.1051/agro/2009019

[26] Gulbe G, Valdovska A. Diversity of microscopic fungi in the raw milk from Latvian organic farms. Proceedings of the Latvia University of Agriculture. 2014;**31**:46–53. doi:10.2478/plua–2014–0006

[27] Skauq MA. Analysis of Norwegian milk and infant formulas for ochratoxin A. Food Additives & Contaminants. 1999;**16**:75–8. doi:10.1080/026520399284235

[28] Tsakiris IN, Tzatzarakis MN, Alegakis AK, Vlachou MI, Renieri EA, Tsatsakis AM. Risk assessment scenarios of children's exposure to aflatoxin M1 residues in different milk types from the Greek market. Food and Chemical Toxicology. 2013;**56**:261–265. doi:10.1016/j.fct.2013.02.024

[29] Schiavone A, Cavallero C, Girotto L, Pozzo L, Antoniazzi S, Cavallarin L. A survey on the occurrence of ochratoxin A in feeds and sera collected in conventional and organic poultry farms in Northern Italy. Italian Journal of Animal Science 2008;**7**:495–503. doi:10.4081/ijas.2008.495

[30] Duarte SC, Lino CM, Pena A. Food safety implications of ochratoxin A in animal-derived food products. The Veterinary Journal. 2012;**192**:286–292. doi:10.1016/j.tvjl.2011.11.002

[31] Winter CK, Davis SF. Organic foods. Journal of Food Science. 2006;**71**:117–124. doi:10.1111/j.1750-3841.2006.00196.x

[32] Rey-Crespo F, Miranda M, Lopez-Alonso M. Essential trace and toxic element concentrations in organic and conventional milk in NW Spain. Food and Chemical Toxicology. 2013;**55**:513–518. doi:10.1016/j.fct.2013.01.040

[33] Chary NS, Kamala CT, Raj DSS. Assessing risk of heavy metals from consuming food grown on sewage irrigated soils and food chain transfer. Ecotoxicology and Environmental Safety. 2008;**69**:513–524. doi:10.1016/j.ecoenv.2007.04.013

[34] McCarty MF, DiNicolantonio JJ. Are organically grown foods safer and more healthful than conventionally grown foods? British Journal of Nutrition. 2014;**112**:1589–1591. doi:10.1017/S0007114514002748

[35] Chandorkar S, Vaze N. Analysis of metal content of organic foods. Journal of Environmental Science. Toxicology and Food Technology. 2013;**4**:44–49.

[36] Gabryszuk M, Sloniewski K, Metera E, Sakowski T. Content of mineral elements in milk and hair of cows from organic farms. Journal of Elementology 2010;**15**:259–267. doi:10.5601/jelem.2010.15.2.259-267

[37] Seğmenoğlu MS. Comparison of some metal levels of organic and conventional milk and dairy products consumed in Turkey [thesis]. Ankara: Ankara University; 2012.

[38] Baydan E, Segmenoglu MS. Some metal levels of organic milk and dairy products consumed in Turkey. In: Livestock Nutrition; 11–12 August 2015; Frankfurt, Germany.

[39] Küçükyılmaz K, Bozkurt M, Yamaner Ç, Çınar M, Çatlı AU, Konak R. Effect of an organic and conventional rearing system on the mineral content of hen eggs. Food Chemistry. 2012;**132**:989–992. doi:10.1016/j.foodchem.2011.11.084

[40] Arslanbaş E, Baydan E. Metal levels in organically and conventionally produced animal and vegetable products in Turkey. Food Additives & Contaminants: Part B. 2013;**6**:130–133. doi:10.1080/19393210.2013.764931

[41] Giannenas I, Nisianakis P, Gavriil A, Kontopidis G, Kyriazakis I. Trace mineral content of conventional, organic and courtyard eggs analysed by inductively coupled plasma mass spectrometry (ICP-MS). Food Chemistry. 2009;**114**:706–711. doi:10.1016/j.foodchem.2008.09.079

[42] Kamal RM, Albadry SM, Bayoumi MA. Toxic metals in organic, home and commercially produced eggs, comparative and risk assessment study. Japanese Journal of Veterinary Research. 2016;**64**:59–64.

[43] Lindén A, Andersson K, Oskasson A. Cadmium in organic and conventional pig production. Archives of Environmental Contamination and Toxicology. 2001;**40**:425–431. doi:10.1007/s002440010193

[44] Bourn D, Prescott J. A comparison of the nutritional value, sensory qualities, and food safety of organically and conventionally produced foods. Critical Reviews in Food Science and Nutrition. 2002;**42**:1–34. doi:10.1080/10408690290825439

[45] Pussemier L, Larondelle Y, Peteghem CV, Huyghebaert A. Chemical safety of conventionally and organically produced foodstuffs: a tentative comparison under Belgian conditions. Food Control. 2006;**17**:14–21. doi:10.1016/j.foodcont.2004.08.003

[46] De Vries M, Kwakkel RP, Kijlstra A. Dioxins in organic eggs: a review. NJAS. 2006;**54**: 207–222.

[47] Woese K, Lange D, Boess C, Bögl KW. A comparison of organically and conventionally grown foods-results of a review of the relevant literature. Journal of Science Food and Agriculture. 1997;**74**:281–293. doi:10.1002/(SICI)1097-0010(199707)74:3<281

[48] Chiesa LM, Labella GF, Giorgi A, Panseri S, Pavlovic R, Bonacci S, Arioli F. The occurrence of pesticides and persistent organic pollutants in Italian organic honeys from different productive areas in relation to potential environmental pollution. Chemosphere. 2016;**154**:482–490. doi:10.1016/j.chemosphere.2016.04.004

[49] Winter CK, Davis SF. Are organic foods healthier. CSA News. 2007;**52**:2–13

[50] Kanbur M, Tekeli MY. Heavy metal residues in meat. Turkiye Klinikleri Journal of Veterinary Sciences Pharmacology Toxicol-Special Topics. 2016;2(1).

[51] Shibamoto T, Bjeldanes LJ. Food Contaminants from Industrial Wastes. In: Introduction to Food Toxicology. Second edition. Academic Press is an imprint of Elsevier, Burlington, USA. 2009; pp. 181–211.

[52] Deshpande SS. Toxic Metals, Radionuclides, and Food Packaging Contaminants. In: Deshpande SS editör. Hand Book of Food Toxicology. Marcel Dekker, Inc., New York, USA, 2002; pp. 783–810.

[53] Omaye ST. Food and nutritional toxicology. New York: CRC Press 2004.

[54] Wuana RA, Okieimen FE. Heavy metals in contaminated soils: A review of sources, chemistry, risks and best available strategies for remediation. ISRN Ecology 2011;1–20. doi:10.5402/2011/402647

[55] Reilly C, Sudarmadji S, Sugiharto E. Mercury and arsenic in soil, water and foods of the Dieng Plateau, Java. In Proceedings of the Nutrition Society. 1989; pp. 14–118.

[56] Türközü D, Şanlıer N. Current overview: Heavy metal contamination of food. Selçuk Tarım ve Gıda Bilimleri Dergisi. 2012;**26**:73–80.

[57] International Agency for Research on Cancer (IARC). IARC Monographs on the Evaluation of Carcinogenic Risks to Humans. Inorganic and Organic Lead Compounds volume 87; 10–17 February 2004; Lyon: France: IARC; 2006.

[58] Baydan E, Yurdakök B. Some contaminants in fish. Veteriner Hekimler Derneği Dergisi. 2010;**81**:77–81.

[59] Erkmen O. Foodborne hazards and safe food production. Çocuk Sağlığı ve Hastalıkları Dergisi 2010;**53**:220–23.

[60] FAO and WHO. Assuring food safety and quality: Guidelines for strengthening national food control systems. Joint FAO/WHO Publication. 2003;1–80.

[61] Dakeishi M, Murata K, Grandjean P. Long-term consequences of arsenic poisoning during infancy due to contaminated milk powder. Environmental Health. 2006;**5**:1–7.

[62] Hasni MJ. Niigata minamata disease. Malaysian Journal of Public Health Medicine. 2014;**14**:47–54.

[63] Ahmad R, Salem NM, Estaitieh H. Occurrence of organochlorine pesticide residues in eggs, chicken and meat in Jordan. Chemosphere. 2010;**78**:667–71. doi:10.1016/j.chemosphere.2009.12.012.

[64] Zastempowska E, Grajewski J, Twaruzek M. Food-borne pathogens and contaminants in raw milk-a review. Annals of Animal Science. 2016;**16**:623–639. doi:10.1515/aoas-2015-0089

[65] Srinivasan V, Sawant AA, Gillespie BE, Headrick SJ, Ceasaris L, Oliver SP. Prevalence of enterotoxin and toxic shock syndrome toxin genes in *Staphylococcus aureus* isolated from milk of cows with mastitis. Foodborne Pathogens and Disease. 2006;**3**:274–83. doi:10.1089/fpd.2006.3.274

[66] Oliver SP, Boor KJ, Murphy SC, Murinda SE. Food safety hazards associated with consumption of raw milk. Foodborne Pathogens and Disease. 2009;**6**:793–806. doi:10.1089=fpd.2009.0302

[67] Altun B, Besler T, Ünal S. Ankara'da satılan sütlerin değerlendirilmesi. Sürekli Tıp Eğitimi Dergisi. 2002;**11**:51–55.

[68] Kuhnen S, Stibuski RB, Honorata LA, Filho LCPM. Farm Management in organic and conventional dairy production systems based on pasture in Southern Brazil and its consequences on production and milk quality. Animals. 2015;**5**:479–494. doi:10.3390/ani5030367

[69] Torkar KG, Teger SG. The microbiological quality of raw milk after introducing the two day's milk collecting system. Acta agriculturae Slovenica. 2008;**92**:61–74.

[70] Becker-Algeri TA, Castagnora D, De Bortoli K, De Souza C, Drunkler DA. Mycotoxins in bovine milk and dairy products: a review. Journal of Food. 2016;**81**:544–552. doi:10.1111/1750-3841.13204

[71] Aidoo KE, Mohamed SM, Candlish AA, Tester RF. Occurrence of fungi and mycotoxins in some commercial baby foods in North Africa. Food and Nutrition Sciences. 2011;**2**:751–758. doi:10.4236/fns.2011.27103

[72] Resanovic, RD, Vucicevic MZ, Trailovic JBN, Maslic-Strizak DN, Jacevic VM. Mycotoxins and their effect on human health. Jour. Nat. Sci, Matica Srpska Novi Sad. 2013;**124**:315–324. doi:10.2298/ZMSPN1324315R

[73] Delavenne E, Mounier J, Asmani K, Le Blay G. Fungal diversity in cow, goat and ewe milk. International Journal of Food Microbiology 2011;**151**:247–51. doi:10.1016/j.ijfoodmicro.2011.08.029

[74] Morshdy AE. MA., Hussien MAM, Elrhem E. A. Aflatoxins Residues in Some Poultry Meat Products. In: 2nd Conference of Food Safety, Suez Canal University, Faculty of Veterinary Medicine Volume I August 2015; Ismailia. Egypt: pp. 96–104

[75] Lau OW, Wong SK. Contamination in food from packaging material. Journal of Chromatography A. 2000;**882**:255–270.

[76] Wu YM, Hu CY, Wang L, Cheng J, Wang ZW, Zhu Y. Migration of Contaminants from Packaging Materials into Dairy Products. Proceedings of the 17th IAPRI World Conference on Packaging. (IAPRI); 12–15 October 2010; Tianjin. China: IAPRI; 2010. pp. 207–211

[77] Ramonaityte DT. Copper, zinc, tin and lead in canned evaporated milk, produced in Lithuania: the initial content and its change at storage. Food Additives & Contaminants. 2001;**18**:31–37. doi:10.1080/02652030010008805

[78] Weisel C, Demak M, Marcus S, Goldstein BD. Soft plastic bread packaging: lead content and reuse by families. American Journal of Public Health. 1991;**81**:756–8.

[79] Kim KC, Park YB, Lee MJ, Kim JB, Huh JW, Kim DH, Lee JB, Kim JC. Levels of heavy metals in candy packages and candies likely to be consumed by small children. Food Research International. 2008;**41**:411–418. doi:10.1016/j.foodres.2008.01.004

[80] Food and Drug Administration (FDA). Lead contamination in candy. FDA consumer. 2004;**38**:7

[81] Khaniki RJ. Chemical contaminants in milk and public health concerns: a review. International Journal of Dairy Science. 2007;**2**:104–115

[82] Blandzic N, Kolanovic BS, Varenina I, Scortichini G, Annunziata L, Brstilo M, Rudan N. Veterinary drug residues determination in raw milk in Croatia. Food Control. 2011;**22**:1941–1948. doi:10.1016/j.foodcont.2011.05.007

[83] Santos JS, Botaro BG. Chloramphenicol residues found in milk processed in Northern Parana, Brazil. Journal für Verbraucherschutz und Lebensmittelsicherheit. 2016;**11**:79–81. doi:10.1007/s00003-015-0975-3

[84] Hussein MA, Khalil S. Screening of some antibiotics and anabolic steroids residues in broiler fillet marketed in El-SharkiaGovernorate. Life Science Journal. 2013;**10**:2011–20118.

[85] Hajrulai-Musliu Z, Uzunov R, Stojkovski V, Dimitrieska-Stojkovik E, Stojanovska-Dimzoska B, Sekulovski P. Determination of Clenbuterol in Meat Samples with ELISA and GC-MS Method. In: 1st Annual International Interdisciplinary Conference (AIIC 2013); 24–26 April 2013; Azores, Portugal: AIIC; 2013. p. 817–825.

[86] Baydan, E., Yurdakök, B. Balıklarda görülen ilaç kalıntıları. Türk Veteriner Hekimleri Birliği Dergisi. 2009;**9**:66–77.

[87] Heshmati A. Impact of cooking procedures on antibacterial drug residues in foods: a review. Journal of Food Quality and Hazards Control. 2015;**2**:33–37.

[88] Baydan E, Filazi A, Kum C, Sekkin S. The effect of cooking and freezing applications on drug residues at the broiler: 2. Effects of cooking and different freezing times on danofloxacine residues. Veteriner Hekimler Derneği Dergisi. 2000;**71**:33–36.

[89] Kanbur M, Liman BC, Eraslan G, Baydan E, Dinç E, Karabacak M. Effects of various cooking and freezing processes on the residues of sulfachlorpyridazine-trimethoprim combination in broiler tissues. Kafkas Üniversitesi Veteriner Fakültesi Dergisi. 2014;**20**:487–492. doi:10.9775/kvfd.2013.10220

[90] Sever E, Baydan E. The Effect of various cooking and freezing processes on the levamisole residues in broiler tissues. Kafkas Üniversitesi Veteriner Fakültesi Dergisi. 2013;**19**: 239–244. doi:10.9775/kvfd.2012.7609

[91] Bajwa U, Sandhu KS. Effect of handling and processing on pesticide residues in food-a review. Journal of Food Science and technology. 2014;**51**:201–220. doi:10.1007/s13197-011-0499-5

[92] Bayyari S, Herrera A, Conchello MP, Arino AA, Lazaro R, Yagüe C. Influence of meat processing and meat starter microorganisms on the degradation of organochlorine contaminants. Journal of Agricultural and Food Chemistry. 1998;**46**:3187–3193. doi:10.1021/jf970932v

[93] Bhuiyan NH, Bhuiyan HR, Ahmed K, Dawla-tana M, Haque F, Rahim M, Bhuiyan NI. Organochlorine insecticides (DDT and heptachlor) in dry fish: traditional washing and cooking effect on dietary intake. Bangladesh Journal of Pharmacology. 2009;**4**:46–50. doi:10.3329/bjp.v4i1.1051

[94] Witczak A. Effect of heat treatment on organochlorine pesticide residues in selected fish species. Polish Journal of Food and Nutrition Sciences. 2009;**59**:231–235.

[95] Commission Regulation. Commission Regulation (EC) No 1881/2006 of 19 December 2006 setting maximum levels for certain contaminants in foodstuffs.

[96] International Federation of Organic Agriculture Movement (IFOAM) [Internet]. 2016. Available from: https://www.ifoam.bio/

[97] EU Agriculture and Rural Development. Organic Farming EU Policy. [Internet]. 2016. Available from: https://ec.europa.eu/agriculture/organic/eu-policy/eu-legislation/brief-overview_en

[98] Codex Alimentarius. Codex Alimentarius Commission, Guidelines For The Production, Processing, Labellingand Marketing of Organically Produced Foods, GL 32–1999.

[99] Sekkin S. Kum C. Possible natural toxins in organic livestock farming. Kafkas Üniversitesi Veteriner Fakültesi Dergisi. 2013;**19**:725–734. doi:10.9775/kvfd.2012.8415

[100] Commission Implementing Regulation (EU) 2016/662 of 1 April 2016 concerning a coordinated multiannual control programme of the Union for 2017, 2018 and 2019 to ensure compliance with maximum residue levels of pesticides and to assess the consumer exposure to pesticide residues in and on food of plant and animal origin.